BEHAVIORAL ECONOMICS AND HEALTHY BEHAVIORS

The field of behavioral economics can tell us a great deal about cognitive bias and unconscious decision-making, challenging the orthodox economic model whereby consumers make rational and informed choices. But it is in the arena of health that it perhaps offers individuals and governments the most value. In this important new book, the most pernicious health issues we face today are examined through a behavioral economics lens. It provides an essential and timely overview of how this growing field of study can reframe and offer solutions to some of the biggest health issues of our age.

The book opens with an overview of the core theoretical concepts, after which each chapter assesses how behavioral economics research and practice can inform public policy across a range of health issues. Including chapters on tobacco, alcohol and drug use, physical activity, dietary intake, cancer screening and sexual health, the book integrates the key insights from the field to both developed and developing nations.

Also asking important ethical questions around paternalism and informed choice, this book will be essential reading for students and researchers across psychology, economics, and business and management, as well as public health professionals wishing for a concise overview of the role that behavioral economics can potentially play in allowing people to live healthier lives.

Yaniv Hanoch is Professor of Decision Science in the School of Psychology, University of Plymouth, UK.

Andrew J. Barnes is Assistant Professor in the Department of Health Behavior and Policy at the Virginia Commonwealth University School of Medicine, Research Associate of Massey Cancer Center, and affiliate faculty in the Center for the Study of Tobacco Products, USA.

Thomas Rice is Professor in the Department of Health Policy and Management, UCLA Fielding School of Public Health, USA, with a joint appointment in Public Policy.

This book is a must-have for those who want to understand how the insights from behavioral economics can be applied to the most significant health issues we face, such as smoking, obesity and prevention of HIV. Each chapter will change the way you think about health behaviors and provide you with up-to-date research distilled to make it accessible. It will be the standard book in behavioral economics and health behaviors for years to come.

Professor Richard Scheffler, *Distinguished Professor of Health Economics and Public Policy, University of California, Berkeley, USA*

BEHAVIORAL ECONOMICS AND HEALTHY BEHAVIORS

Key Concepts and Current Research

Edited by Yaniv Hanoch, Andrew J. Barnes, and Thomas Rice

Routledge
Taylor & Francis Group
LONDON AND NEW YORK

First published 2017
by Routledge
2 Park Square, Milton Park, Abingdon, Oxon OX14 4RN

and by Routledge
711 Third Avenue, New York, NY 10017

Routledge is an imprint of the Taylor & Francis Group, an informa business

Library of Congress Cataloging-in-Publication Data
A catalog record for this title has been requested

ISBN: 978-1-138-63820-4 (hbk)
ISBN: 978-1-138-63821-1 (pbk)
ISBN: 978-1-315-63793-8 (ebk)

Typeset in Bembo
by codeMantra

MIX
Paper from responsible sources
FSC™ C013985

Printed in the United Kingdom
by Henry Ling Limited

CONTENTS

ACKNOWLEDGMENTS

Editing a book requires the efforts of many authors. We would first like to thank the authors of this edited volume for their dedication, hard work, and timely submission of their respective chapters. Their efforts and willingness to read and comment on others' chapters is also greatly appreciated. Their contribution to the completion of this project was extremely valuable. Several external reviewers were kind enough to read and provide excellent comments on various chapters. In particular we would like to thank Fred Zimmerman from UCLA, who was supportive of our project from its early stages. He provided detailed comments and suggestions about the book proposal as well as on various chapters. Jessica Greene, from George Washington University, and Chao Zhou, from the U.S. Centers for Disease Control and Prevention, read our chapter on health insurance and gave us excellent suggestions. Yaniv Hanoch would like to thank Michaela Gummerum for her ongoing support and excellent ideas; Tom Rice expresses his great appreciation for the continued advice and support from Kate Desmond; and Andrew Barnes would like to thank Kate and Ambrose Barnes for letting him work on this project that brought him so much joy over weekends, holidays, and vacations.

ABOUT THE EDITORS AND AUTHORS

Editors

Yaniv Hanoch is Professor of Decision Science in the School of Psychology, University of Plymouth, UK. Professor Hanoch is interested in the intersection between decision science, health economics, and psychology. His research interests include consumer decision-making (especially with regard to health insurance), communicating (health) risk information, medical decision-making, offenders' decision-making and risk-taking, and life-span changes in risk-taking. He is currently serving as an associate editor of the *Journal of Behavioral and Experimental Economics*.

Andrew J. Barnes is Assistant Professor in the Department of Health Behavior and Policy at the Virginia Commonwealth University School of Medicine, Research Associate of Massey Cancer Center, and affiliate faculty in the Center for the Study of Tobacco Products. His training is in health policy and economics and his research interests include applying behavioral economics to health policies, particularly in the areas of substance use and health insurance. Dr. Barnes is the co-author of the book *Healthcare Systems in Transition: United States of America*.

Thomas Rice is Professor in the Department of Health Policy and Management, UCLA Fielding School of Public Health, with a joint appointment in Public Policy. He is a health economist who has studied national health care systems, competition and regulation, behavioral economics, physicians' economic behavior, health insurance, and the Medicare program. The fourth edition of his book, *The Economics of Health Reconsidered*, was published in 2016. He led a team of researchers that wrote a book published in 2013 about the US health care system, for the European Observatory on Health Systems and Policies. Dr. Rice served as editor of the journal, *Medical Care Research and Review*, from 1994 to 2000.

Authors

Michael Amlung is an Assistant Professor in the Department of Psychiatry & Behavioural Neurosciences in the Michael G. DeGroote School of Medicine at McMaster University, Ontario, where he directs the Behavioural Sciences Core of the Peter Boris Centre for Addictions Research. A cognitive neuroscientist by training, his research interests include applying behavioral economics and neuroeconomics principles to understand the etiology and treatment of addictive disorders.

Warren Bickel is the Director of the Addiction Recovery Research Center at the Virginia Tech Carilion Research Institute and Virginia Tech Carilion Professor of Behavioral Health Research. Dr. Bickel's research examines the decision-making processes underlying dysfunctional behaviors such as addiction and other poor health behaviors. Having co-edited five books and published over 350 papers and chapters, Dr. Bickel's work is frequently cited and receives national and international recognition.

Marie A. Bragg is an Assistant Professor in the Section on Health Choice, Policy and Evaluation at the NYU School of Medicine, with a joint faculty appointment at the NYU Global Institute of Public Health. A clinical psychologist by training, Dr. Bragg conducts research on environmental and social factors associated with obesity, food marketing, food policy, and health disparities.

Jason N. Doctor is Director of Health Informatics at the Leonard D. Schaeffer Center for Health Policy and Economics and Associate Professor in the Department of Pharmaceutical and Health Economics, at the University of Southern California School of Pharmacy. A health psychologist by training, his research interests include using behavioral economics to improve the quality of care in medicine.

Brian Elbel is an Associate Professor of Population Health and Health Policy within the Department of Population Health at the NYU School of Medicine, with a joint faculty appointment at the NYU Wagner Graduate School of Public Service. Trained in health policy/health economics, Dr. Elbel studies how individuals make decisions that influence their health, with a particular emphasis on behavioral economics, evaluation, obesity, and food choice.

Richard Frank is the Margaret T. Morris Professor of Health Economics in the Department of Health Care Policy at Harvard Medical School. He has conducted research on how behavioral economics can apply to health insurance arrangements, physician payment systems, and mental health and substance use disorder policy. From 2014–2016 he served as Assistant Secretary for Planning and Evaluation at the U.S. Department of Health and Human Services.

Joshua Gray is a doctoral student in the Clinical Psychology Program at the University of Georgia. His research seeks to elucidate the neurobiological underpinnings of risk phenotypes for addiction to better prevent and treat addictive disorders. Josh has used behavioral economics, neuroimaging, and molecular genetics methodologies to better understand addictive processes.

Tammy Leonard is Associate Professor of Economics at the University of Dallas. She specializes in interdisciplinary applications of public, urban and behavioral economics along with applied spatial and econometric analysis methods. Dr. Leonard is also co-director of the Community Assistant Research (CARE) initiative, which leverages interdisciplinary relationships between academic researchers and community stakeholders to improve research related to low-income households.

Yan Li is a Research Scientist at the Center for Health Innovation, The New York Academy of Medicine, and an Assistant Professor in the Department of Population Health Science and Policy at the Icahn School of Medicine at Mount Sinai. A biomedical and systems engineer by training, his research interests include simulation modeling, cost-effectiveness analysis, behavioral economics and social determinants of health. Working with interdisciplinary teams, he has developed a range of innovative computer simulation models for chronic health conditions such as cardiovascular disease, diabetes, and cervical cancer.

Sebastian Linnemayr is a Senior Economist at the RAND Corporation in Santa Monica. An economist by training, his research interests include the design of incentives for long-term health behavior change. Dr. Linnemayr is Principal Investigator on several NIH-funded grants in Uganda using behavioral economics to improve medication adherence of clients in HIV care.

Jill Luoto is an Economist at RAND, a non-profit policy research organization. An economist by training, her research interests include labor, health and behavioral economics, with a focus on poverty and individual decision-making. Much of her work has focused on developing country populations.

James MacKillop is the Peter Boris Chair in Addictions Research and Professor in the Department of Psychiatry and Behavioural Neurosciences at McMaster University. A clinical psychologist by training, he conducts translational research on addictive behavior, especially the application of behavioral economics and neuroeconomics, to understand alcohol use disorder, nicotine dependence and other addictive disorders.

Daniella Meeker is an Assistant Professor at the University of Southern California (USC) Keck School of Medicine and an Information Scientist at RAND. She directs the Informatics Program for the Southern California Clinical Translational Sciences Institute, a collaboration between Children's Hospital of Los Angeles, Los Angeles County Department of Health Services, and Keck Medicine of USC.

Alexandra Mellis is a graduate student in the Translational Biology, Medicine, and Health Ph.D. program at Virginia Tech. Her research interests include the impact of narratives on health behavior and decision-making.

Steven Meredith is a postdoctoral fellow at the Calhoun Cardiology Center at the University of Connecticut School of Medicine. A behavioral pharmacologist by training, his research interests include behavioral economics interventions to treat substance abuse and other behavioral health problems.

Lara N. Moody is a clinical psychology doctoral student at Virginia Tech. Her research interests include improving treatments for substances of abuse, with a particular interest in providing improved treatments to underserved populations.

José A. Pagán is Director of the Center for Health Innovation at The New York Academy of Medicine and Professor in the Department of Population Health Science and Policy at the Icahn School of Medicine at Mount Sinai. He is also Adjunct Senior Fellow of the Leonard Davis Institute of Health Economics at the University of Pennsylvania. His research interests include systems science, health disparities and population health management.

Nancy Petry is Professor of Medicine, and Director of Behavioral Cardiology Prevention and the REWARD Center at the Calhoun Cardiology Center at the University of Connecticut School of Medicine. A psychologist by training, her research interests include behavioral therapies for treatment of addictive disorders ranging from substance use to gambling disorders. Her work on improving adherence behaviors has extended to diabetes management, weight loss, exercise, and medication adherence.

Amanda J. Quisenberry is a Postdoctoral Associate at the Addiction Recovery Research Center of the Virginia Tech Carilion Research Institute. Dr. Quisenberry's training and research interests include addiction, recovery, behavioral pharmacology, and behavior analysis.

Aditi P. Sen is an Assistant Professor in the Department of Health Policy and Management at the Johns Hopkins Bloomberg School of Public Health. An economist by training, her research interests include how providers and payers interact in health care markets and how behavioral economics can be applied to provider and consumer behavior. From 2015–2016, she was a Health and Aging Policy Fellow in the office of the Assistant Secretary for Planning and Evaluation at the U.S. Department of Health and Human Services.

Kerem Shuval is the Director of Physical Activity and Nutrition Research in the Economic and Health Policy Research Program, Department of Intramural Research, American Cancer Society. Trained in health behavior change, evidence-based medicine, and health economics, his research aims to better understand decision-making. That is, why some individuals make healthier choices while others engage in self-harming behaviors.

Sarah E. Snider is a Post Doctoral Associate at the Addiction Recovery Research Center as part of the Virginia Tech Carilion Research Institute. A behavioral pharmacologist and toxicologist by training, her research interests include drug use behavior, decision-making, and candidate treatments for substance use disorder.

Jeff S. Stein is a research assistant professor in the Addiction Recovery Research Center at the Virginia Tech Carilion Research Institute. A behavioral economist by training, his research interests include tobacco product abuse liability and the etiology and treatment of addictive disorders.

Roberto Villarreal is Senior Vice President for Research and Information Management at University Health System in San Antonio, Texas and Associate Professor in the Department of Family and Community Medicine at The University of Texas Health Science Center at San Antonio. He is a physician interested in health promotion and disease prevention related to the implementation and evaluation of community intervention programs. During the past 20 years, Dr. Villarreal has participated in the development of trans-theoretical models that have been applied in cancer prevention, diabetes, and injury prevention and control.

Fernando A. Wilson is Associate Professor in the Department of Health Services Research and Administration at the College of Public Health, University of Nebraska Medical Center. He is also Acting Director of the Center for Health Policy at the University of Nebraska Medical Center and his research interests include health policy and services, health economics, traffic safety, immigrant health, and access to care.

PART I

Background material

1

INTRODUCTION

Yaniv Hanoch, Andrew J. Barnes, and Thomas Rice

Tackling poor health behaviors

One of the biggest challenges facing governments around the world is improving people's health while simultaneously controlling health expenditures that are now responsible for costs amounting to around US$7.2 trillion per year (World Health Organization [WHO], 2014). Multiple factors contribute to poor health, such as the environment, governmental policies, access to health care, and genes. Personal choices or behaviors have been identified as primary contributors to people's poor health. Indeed, many health care interventions are specifically designed to improve unhealthy behaviors, such as substance abuse, poor diet, and lack of physical activity. According to the Centers for Disease Control and Prevention (Yoon et al., 2014), up to 40% of deaths from the five leading causes are preventable, as they relate to unhealthy behaviors.

Personal behaviors and choices influence a wide range of conditions. Cigarette smoking, one of the leading causes of premature death, is still highly prevalent in countries such as China, where 68% of males are smokers and an estimated 1 million people die annually from tobacco use (Chen et al., 2015). According to the *Global Status Report on Alcohol and Health* (World Health Organization, 2014b), a similar trend has emerged for alcohol consumption: alcohol misuse is associated with 3.1 million deaths a year. In a recent study, Gowing and colleagues (2015) estimated that just under 5% of the world population can be classified as having an alcohol use disorder. Drug abuse is no different: It is estimated that 27 million adults worldwide are problem drug users, with 0.5% reported using cocaine- and amphetamine-type drugs (Gowing et al., 2015; UNODC, 2012). In 2014, over 34 million people were living with HIV (with about 2 million being newly infected), with over 25 million of them in Africa (World Health Organization, 2016a)—mostly as a result of unprotected sex or intravenous drug use. Another line of research has shown that even health care providers are not immune to making bad health decisions, such as not following recommended medical procedures and guidelines, and prescribing antibiotics to patients with a common cold (Harris et al., 2016).

Obesity rates (body mass index greater than or equal to 30.0) in the United States, likewise, increased from 22.9% during the years from 1984 to 1994, to 36.3% in the period between 2011 and 2014 (Fryar et al., 2014). Across the globe, the trend has been just as alarming: In 2014 more than 1.6 billion people were overweight, and 600 million of them were classified as obese (World Health Organization, 2016b), doubling the rate of obesity since 1980. In response to the growing concern over obesity, the WHO (2000) published a report on preventing and managing the phenomenon. Among its conclusions, one point is especially pertinent to this book: "Obesity is a serious disease, but its development is not inevitable. It is largely preventable through lifestyle changes" (p. 4).

Although the WHO statement referred specifically to obesity, it clearly applies to many health behaviors affecting morbidity and mortality, such as smoking, drug abuse, lack of physical activity, and poor diet. Furthermore, one of the key messages that the WHO report is that many behaviors are amenable to change. Indeed, smoking, taking drugs, not exercising, drinking alcohol, choosing and sticking with low-value health plans, and mis-prescribing drugs are a few examples of behaviors that can be changed to improve health outcomes. The more difficult and fundamental question, however, is how people can be convinced to change behaviors. For example, what can be done to reduce smoking, drug abuse, and alcohol consumption? What can be done to increase exercising rates and duration? How can we improve health care plan choices? And how can we improve physicians' antibiotic prescriptions?

Economic solutions: traditional and behavioral

It should be clear from the magnitude of the health problems described above that no single antidote will cure all these complex problems. Traditional economic approaches have had some success in modifying behavior. For example, an increase in cigarette and alcohol prices has been associated with reductions in the consumption of these substances, but high smoking rates persist in many countries and across a number of subpopulations in the United States. Providing consumers with information about health insurance plans can aid them in making better choices (Barnes et al., in press), but despite these efforts, the rate of switching to more cost-effective plans has been lower than what economic theory would predict.

The approaches employed in traditional economics to alter behavior, such as pricing and information-based strategies, represent a critically important set of tools for regulators and governments focused on improving health and health care. However, growing literature has pointed out the shortcomings of traditional economic thinking and ideas, as well as their somewhat limited success achieved in changing behavior. For example, the idea of *Homo economicus*, that is, rational economic man, has been shown to be problematic. Herbert Simon's (1955, 1956) introduction of the term *bounded rationality* was one early attempt to highlight the shortcomings of traditional theory and, since then, a host of psychologists and economists have provided empirical evidence that further calls it into question.

Perhaps the most important challenge to the traditional economic theory of individual behavior has been the development of behavioral economics. Incorporating insights from psychology and neuroscience, behavioral economics diverges from traditional economics in that it does not assume that agents are fully rational, or make decisions that always maximize their expected utility. Rather, it works from the assumption that agents are limited in their computational abilities, do not possess full information, lack perfect willpower, make decisions that are often affected by trivial differences in their environment, and frequently make choices that deviate from their best self-interest. Working from within this framework, behavioral economics has already made promising contributions in the domain of health behaviors. Indeed, behavioral economics offers rich and novel insights into a spectrum of old, persistent, and complex health-related problems. Tackling these problems can help reduce costs across the globe, improve people's health and well-being, and allow people to make better decisions.

The complex nature of changing health behavior, and the high price (both financial and related to personal well-being) associated with poor health, served as a partial motivation to develop this book. The need to advance new methods to tackle these complex behaviors was another. Behavioral economics offers one promising line of reasoning and its insights can supplement existing approaches. In fact, a number of governments have already taken advantage of the insights from behavioral economics in developing and designing policies. The U.K. government, one of the pioneers in the field, established the Behavioral Insight Team (sometimes dubbed the "Nudge Unit") in 2010 to examine ways that behavioral economics could help tackle a host of policy problems, among them health behaviors. A few years later, the Social and Behavioral Sciences Team (SBST) was inaugurated in the United States. Early SBST projects include improving registration for the Federal Health Insurance Marketplace and increasing flu vaccination rates. The Behavioral Insight Team and the SBST are two examples where behavioral economics has injected a novel perspective.

This book offers a window into the opportunities and challenges that behavioral economics offers to address a wide spectrum of health behaviors. Needless to say, no single book can cover the entire range of health problems that can potentially be addressed with behavioral economics. Furthermore, given the relatively recent development of behavioral economics, its ideas and promises have not been tested in many health-related areas. Thus, the book should serve as an inspiration and a guide to the type of approaches employed thus far.

Organization of the book

The chapters in this book tackle issues on both the individual and government level, and they range from personal behavior to government policies. The book is divided into three broad sections: Part II: Shaping Health Behaviors, Part III: Detecting and Managing Disease, and Part IV: The Role of Providers, Insurers, and Government. Before Part II, however, Chapter 2 provides readers with a brief overview of behavioral economics. A solid understanding and knowledge of the underlying principles governing economics and specifically behavioral economics are essential for making use of the entire book and for those wishing to develop these ideas further.

Part II: Shaping health behaviors

Smoking represents one of the greatest public health problems. In fact, the WHO argued that smoking "is one of the biggest public health threats the world has ever faced," with over 5 million deaths per year (WHO, 2016c). Reducing tobacco use, hence, has the potential to reduce morbidity and mortality rates worldwide. There is little doubt that using traditional economic approaches, particularly increasing prices (taxes), has led to a reduction in tobacco use. Yet, advances over the past three decades have provided us with additional innovative means to tackle this important public health problem.

In Chapter 3, by Warren K. Bickel, Lara N. Moody, Sarah E. Snider, Alexandra M. Mellis, Jeffrey S. Stein, and Amanda J. Quisenberry, the authors review four behavioral economics techniques—operant self-administration, hypothetical purchase task, naturalistic demand assessment, and experimental tobacco marketplace—that have made a substantial contribution to our knowledge about tobacco use and addiction. In the chapter, the authors argue that while traditional economic tools have been useful in informing us about historical trends, employing behavioral economics tools, both in and outside the lab, can provide more up-to-date evidence. Operant self-administration—a method that allows researchers to examine the effects of price on tobacco self-administration in the lab—has afforded researchers important insights on how price affects tobacco use and thus how it might affect smokers' purchasing behavior. Hypothetical purchasing measures, where individuals are asked how much tobacco product they would purchase at varying prices, have allowed investigators to capture purchasing behavior using a technique that is cheaper and more efficient to employ than traditional measures. Naturalistic demand assessment builds on hypothetical purchasing measures, but with the important extension of collecting real-world data, both with regard to price change as well as actually giving the tobacco products to participants. Naturalistic demand assessment, thus, can be important in substantiating and validating laboratory findings. Finally, experimental tobacco marketplaces have allowed researchers to develop a rigorous study protocol and carefully manipulate variables of interest (i.e., product, price, brand name, strength, flavor, etc.) to evaluate their possible effects on behavior. Chapter 1, thus, provides policymakers with insights into how different policies might affect tobacco consumption and gives researchers a spectrum of tools to further investigate alternative methods for reducing tobacco use.

According to the WHO (2015), alcohol misuse is responsible for 3.3 million deaths a year worldwide (or 5.9% of all deaths), and illicit drug use accounts for another 200,000 (UNODC, World Drug Report, 2012). With enormous financial, health, and social ramifications, reducing alcohol and drug use has long been of interest to researchers and policymakers alike. Yet, there is still no consensus on the factors associated with drug and alcohol misuse, nor on how best to prevent and treat these disorders. Traditional economics has relied on price elasticity of demand (imposing taxes or setting a minimum price per unit) and market regulation (prohibiting the sale of alcohol to people under 18) as two key approaches to battling misuse of alcohol and illicit drug use. Behavioral economists, on the other hand, have focused on the notion of delay discounting—or the tendency to place a greater value on immediate versus future rewards—in their attempt to address the problem. They have also developed more sophisticated tools that can better capture demand. Chapter 4, by Michael Amlung, Joshua Gray, and James MacKillop, provides an overview of the approaches taken in behavioral economics to gain a better understanding of the mechanisms underlying addictive behavior, and delineates clinical methodologies for preventing and treating addiction. Among the techniques designed to alter delay-discounting rates and engagement with alcohol are episodic future thinking (EFT)—one that requires participants to project themselves into the future in order to pre-experience the event, and substance-free activity sessions (SFAS)—a method designed to increase the salience of the person's future goals, highlight the potentially negative association between substance use and goal achievement, and increase engagement in substance-free alternative activities.

Along with stopping smoking, and reducing alcohol intake, increasing physical activity is one of the most common pieces of health advice provided by public health authorities. Indeed, according to the Centers for Disease Control and Prevention, physical activity can help reduce the risk of cardiovascular disease, type 2 diabetes, metabolic syndrome, and some cancers. It can also improve mental health, mood and the chances of living longer, enhance the ability to do daily activities, prevent falls, as well as help control weight and strengthen bones and muscles. Despite the host of benefits linked to physical activity, relatively few adults (Troiano et al., 2008) adhere to the recommendations put forth by health authorities (e.g., the American Heart Association recommends 30 minutes of moderate-intensity aerobic activity at least 5 days per week for a total of 150 minutes). Early interventions focused on the individual level, with the principal idea being that individuals make rational decisions based on the costs and benefits associated with engaging in physical activity. Some researchers have come to realize that a multi-level approach that incorporates the individual, social/cultural, organizational, environmental, and policy levels would be more conducive to improving physical activity levels (Owen et al., 2011; Sallis et al., 2012). Chapter 5, by Tammy Leonard and Kerem Shuval, reviews a host of measures that can be used on both the individual and organizational level to encourage physical activity. These include supporting physical activity routines at work, designing environments that naturally boost physical activity (such as playgrounds), offering incentives based on objective measures (such as the length of time exercised), establishing pre-commitment schemes, and framing physical activity messages in a positive light (rather than emphasizing the negative consequences). While more data is needed to evaluate the merits of behavioral economics in improving (long-term) physical activity rates, early results are promising.

Another health-related behavior that has garnered much attention is diet. The U.S. Department of Health and Human Services and the U.S. Department of Agriculture (2015) dietary guidelines for 2015–2020 contain five overarching recommendations for consumers: follow a healthy eating pattern across the life span; focus on variety, nutrient density, and amount; limit calories from added sugars and saturated fats, and reduce sodium intake; shift to healthier food and beverage choices; and support healthy eating patterns for all. Consumers' choices and behaviors are, of course, the crucial ingredients for adhering to these guidelines. Despite the ongoing publication of dietary recommendations, the rate of obesity in the United States (and in many countries around the world) has doubled since the early 1970s. Chapter 6, by Marie A. Bragg and Brian Elbel, first argues that a large corpus of data brings into question the utility of interventions based on educational campaigns (such as providing calorie information) that assume consumers will make rational decisions based on the available information. Rather, they argue that, in additional to traditional economics measures such as taxation, there is a need to focus on a range of environmental factors—such as access to playgrounds and fresh food, food prices—that play a crucial role in consumers' dietary habits. The chapter presents several ideas inspired by behavioral economics, such as proposals for changing the ratio of soft drinks (low- or no-calorie beverages vs. high-calorie drinks) in vending machines, offering easier and faster checkout for those ordering healthy food in fast food places, offering healthy choices as the default option, and altering the food products offered at schools and hospitals.

Part III: Detecting and managing disease

Having first provided evidence on how our biases can shape health behaviors in earlier chapters, the second section of this book synthesizes applications of behavioral economics theory to improve the detection and management of chronic diseases. Almost half of Americans take a prescription drug (Centers for Disease Control and Prevention, 2015a) and non-adherence is commonplace, resulting in substantial costs to individuals and society (Osterberg and Blaschke, 2005). A recent Cochrane review of medication adherence interventions suggests a preponderance of the interventions to increase adherence, many of which rely on often complex combinations of education and peer support, but which are limited in their effectiveness (Nieuwlaat et al., 2014). Chapter 7, by Steven E. Meredith and Nancy M. Petry, examines behavioral economics approaches to increasing medication adherence. The authors focus on what they consider to be a simpler tack: reducing financial barriers and incentivizing adherence. Meredith and Petry review behaviorally informed interventions to reduce non-adherence, particularly interventions that provide proximate reinforcers like small financial incentives when patients take a dose of medication as prescribed. The authors conclude that a variety of incentives informed by behavioral economics can be employed to improve medication adherence across diverse populations and settings.

Although cancer survival rates are improving as a result of advances in cancer screening and treatment technologies, these gains are not equitable, and substantial disparities in cancer outcomes persist (Siegel et al., 2014; Smith et al., 2014). In Chapter 8, Yan Li, Fernando A. Wilson, Roberto Villarreal, and José A. Pagán document the implementation of two such programs designed to improve colorectal and cervical cancer screening for Hispanic adults. In the first program, which targets colorectal cancer screening in Hispanic men, the authors apply behavioral economics insights, including how social and cultural norms influence treatment seeking, to a patient navigation program. The second cancer screening intervention presented also adapts patient navigation programs to incorporate behavioral insights to increase cervical cancer screening in Hispanic women. They find that both behaviorally informed navigation programs increase cancer screening uptake and these increases could lead to improvements in quality of life. In addition to improving colorectal and cervical cancer screening uptake, the authors find the behaviorally informed navigation programs they examined were also cost-effective. Taken together, this chapter provides support for culturally tailored interventions incorporating principles from behavioral economics as promising solutions to reducing disparities in access to, and benefits derived from, cancer detection and treatment.

Nearly 25 million people are living with HIV in sub-Saharan Africa. In the US, more than 1.2 million are living with the disease, with more than 1 in 8 infected persons unaware they carry the virus (AVERT, 2015; Centers for Disease Control and Prevention, 2015b). In Chapter 9, Sebastian Linnemayr argues that many previous interventions to mitigate HIV transmission and improve the quality of life of those infected have struggled to improve behavior in a sustainable and cost-effective manner. These challenges arise in part from a failure to incorporate the biases that shape risk-seeking behavior and engagement with prevention and treatment. The chapter follows the treatment pathway from prevention, HIV testing, linkage to care, to adherence to antiretroviral medication and viral suppression, contrasting approaches from traditional and behavioral economics to support behavior change. The evidence on the effectiveness of interventions based on traditional economic theory reviewed in the chapter is mixed, suggesting such approaches offer limited insight as to the mechanisms contributing to success or failure. However, interventions targeting HIV prevention and treatment that leverage behavioral insights about how, when, and why financial rewards work can be incorporated into programs to make them more effective. In addition to using behavioral economics theory to influence the HIV treatment cascade, behavioral insights can also be applied to messaging in interventions. The chapter discusses the promise of interventions using mobile health (mHealth) platforms, and how such programs represent the next generation approaches leveraging behavioral economics to combat the HIV epidemic in the US as well as developing countries.

There is little doubt that people living in resource-constrained settings confront myriad obstacles to improving their health. Moreover, the judgment and decision-making biases people face, regardless of income or country, compound these barriers. Chapter 10, by Jill Luoto, surveys the literature on how behavioral insights have been applied to improve the detection and management of diseases among developing country populations. This chapter reviews a broad array of evidence on behavioral economic-based interventions to improve health in developing countries. The interventions discussed include completing recommended medical visits, adoption of insecticide-treated bed nets or other preventive health products, and smoking cessation. Luoto finds the growth in the application of behavioral economics interventions in developing countries is occurring for two reasons. First, those living in developing countries make more decisions with health consequences (e.g., safety of drinking water, adequacy of sanitation). Second, the proliferation of behavioral economics research has coincided with that from development economics, with both fields promulgating randomized field experiments to test the interventions that affect health behaviors. The chapter presents important evidence that behaviorally informed interventions can offer a significant advantage in scarce resource settings by designing more efficient policies that better reflect how humans actually behave rather than how traditional theory says they ought to.

Part IV: The role of providers, insurers, and government

The second and third sections of the book focused specifically on using behavioral economics to improve the health behaviors of individuals. Each of the chapters contained recommendations for improving these behaviors but these recommendations were, for the most part, not directly related to providers, insurers, or government. This section of the book aims to fill this gap.

Indeed, consumer behavior is the main focus of behavioral economics. It should be no surprise, however, that health professionals like physicians are subject to some of the same cognitive biases as are consumers. They may, for example, stubbornly stick with past medical practices even when changing them could improve the health of their patients (status quo bias). Or they may order unnecessary procedures because they believe that their patients want something *done* during or after a visit (action bias). Chapter 11, by Daniella Meeker and Jason N. Doctor, focuses on ways in which behavioral economics can be used to improve the clinical quality of physician care. The authors focus on strategies aimed at reducing the inappropriate prescribing of antibiotics. Some of the promising strategies discussed include: public commitments, where, in one study, inappropriate prescribing fell by almost 20 percentage points when the physician put up a poster in the waiting room indicating her pledge to prescribe antibiotics judiciously; peer comparisons, where ranking and then communicating physicians' rates of inappropriate prescribing almost eliminated the problem in one experiment; and accountable justification, by which physicians must provide a written justification in the medical record—which can be viewed by other physicians—when they do not follow practice guidelines in antibiotic prescribing; in one study this reduced inappropriate behavior by about 75%.

Health insurance has also been an area that has received a good deal of attention by behavioral economists. Study after study has shown that people do not make the best choices for themselves when choosing insurance policies. They are hobbled by such things as terminology that they do not understand, poorly organized information, and oftentimes a bewildering amount of choice. Sometimes they choose policies that are dominated in all dimensions by other choices (Bhargava et al., 2015; Sinaiko and Hirth, 2011); in other cases, they simply leave money on the table by not making an optimal choice (Abaluck and Gruber, 2013; Zhou and Zhang, 2012). Chapter 12, authored by the book's editors, Andrew J. Barnes, Thomas Rice, and Yaniv Hanoch, discusses behavioral economics strategies that policymakers can use to facilitate good decision-making. Under one strategy, called *smart defaults*, an employer or government would enroll a person into a health insurance plan that best fits their circumstances, with the proviso that the person could opt for another choice if she liked. Other strategies discussed in the chapter include providing *just in time education* so that consumers have the most relevant information at hand when making decisions, and the use of *choice architecture*, where information is winnowed down to emphasize the most critically important points for good decision-making—highlighting it so as to facilitate comparisons of alternative choices.

The role of public policy is front and center in behavioral economics. The types of policies that typically are recommended based on traditional economic theory are usually limited to tweaking prices or providing more information. In contrast, behavioral economics offers a much richer menu. In Chapter 13, the last chapter of the book, Aditi P. Sen and Richard G. Frank tackle the issue of how government can be used to improve welfare through the use of behavioral economics tools. They posit that the key to appropriate government action is the ability to predict how people will behave. Behavioral economics tools can be used to anticipate this behavior, taking into account people's non-standard beliefs, decision-making, and preferences, as well as social influences such as peer effects. The chapter then provides a number of examples of how behavioral economics can improve traditional policies, and even more importantly, new policy tools that can be derived from an understanding of behavioral economics. By closing the gap between what people truly want versus what they actually consume, the authors argue that behaviorally-based government policies can provide a "richer understanding of individual behavior [that] can then be used to design more effective policy with the aim of promoting healthy behaviors and overall well-being."

References

Abaluck, J. and Gruber, G. (2013). "Evolving Choice Inconsistencies in Choice of Prescription Drug Insurance," National Bureau of Economic Research Working Paper 19163. Available at http://www.nber.org/papers/w19163.pdf (accessed December 31, 2016).

AVERT. (2015). "HIV and AIDS in Sub-Saharan Africa Regional Overview," *Regional Overview*. Available at http://www.avert.org/professionals/hiv-around-world/sub-saharan-africa/overview (accessed December 31, 2016).

Barnes, A., Hanoch, Y., Rice, T., and Long, S. (in press). "Moving Beyond Blind Men and Elephants: Providing Total Estimated Annual Costs Improves Health Insurance Decision-making," *Medical Care Research and Review*.

Bhargava, S., Loewenstein, G., and Sydnor, J. (2015). "Do Individuals Make Sensible Health Insurance Decisions? Evidence from a Menu with Dominated Options," Working Paper 21160, National Bureau of Economic Research. Available at http://www.nber.org/papers/w21160 (accessed December 31, 2016).

Centers for Disease Control and Prevention (2015a). "Therapeutic Drug Use." Available at http://www.cdc.gov/nchs/fastats/drug-use-therapeutic.htm (accessed 29 January 2016).

Centers for Disease Control and Prevention. (2015b). "HIV in the United States: At a Glance," *Statistics Center*. Available at http://www.cdc.gov/hiv/statistics/overview/ataglance.html (accessed December 31, 2016).

Chen, Z., Peto, R., Zhou, M., Iona, A., Smith, M., Yang, L., et al. (2015). "Contrasting Male and Female Trends in Tobacco-Attributed Mortality in China: Evidence from Successive Nationwide Prospective Cohort Studies," *Lancet* 386: 1447–1456.

Fryar, C.D., Carroll, M.D., and Ogden, C.L. (2014). "Prevalence of Overweight, Obesity, and Extreme Obesity Among Adults: United States, 1960–1962 Through 2011–2012," Available at http://www.cdc.gov/nchs/data/hestat/obesity_adult_11_12/obesity_adult_11_12.htm (accessed August 1, 2016).

Gowing, L.R., Ali, R.L., Allsop, S., Marsden, J., Turf, E.E., West, R., et al. (2015). Global Statistics on Addictive Behaviours: 2014 Status Report," *Addiction* 110, 904–919.

Harris, A.M., Hicks, L.A., and Qaseem, A., (2016). "Appropriate Antibiotic Use for Acute Respiratory Tract Infection in Adults: Advice for High-Value Care from the American College of Physicians and the Centers for Disease Control and Prevention," *Annals of Internal Medicine* 64(6): 425–434.

Nieuwlaat, R., Wilczynski, N., Navarro, T., Hobson, N., Jeffery, R., Keepanasseril, A., et al. (2014). "Interventions for Enhancing Medication Adherence," *Cochrane Database of Systematic Reviews* 11.

Osterberg, L. and Blaschke, T. (2005). "Adherence to Medication," *New England Journal of Medicine* 353(5): 487–497.

Owen, N., Sugiyama, T., Eakin, E.E., Gardiner, P.A., Tremblay, M.S., and Sallis, J.F. (2011). "Adults' Sedentary Behavior Determinants and Interventions," *American Journal of Preventive Medicine* 41(2): 189–196.

Sallis, J.F., Floyd, M.F., Rodríguez, D.A., and Saelens, B.E. (2012). "Role of Built Environments in Physical Activity, Obesity, and Cardiovascular Disease," *Circulation* 125(5): 729–737.

Siegel, R., Ma, J., Zou, Z., and Jemal, A. (2014). "Cancer Statistics, 2014," *CA: A Cancer Journal for Clinicians* 64(1): 9–29.

Simon, H.A. (1955). "A Behavioral Model of Rational Choice," *Quarterly Journal of Economics* 69: 99–118.

Simon, H.A. (1956). "Rational Choice, and the Structure of the Environment," *Psychological Review* 63: 129–138.

Sinaiko, A.D. and Hirth, R.A. (2011). "Consumers, Health Insurance and Dominated Choices," *Journal of Health Economics* 30(2): 450–457.

Smith, R.A., Manassaram-Baptiste, D., Brooks, D., Cokkinides, V., Doroshenk, M., et al. (2014). "Cancer Screening in the United States, 2014: A Review Of Current American Cancer Society Guidelines And Current Issues In Cancer Screening," *CA: A Cancer Journal for Clinicians* 64(1): 30–51.

Troiano, R.P., Berrigan, D., Dodd, K.W., Masse, L.C., Tilert, T., and McDowell, M. (2008). "Physical Activity in the United States Measured by Accelerometer," *Medicine and Science in Sports and Exercise* 40: 181–188.

United Nations Office on Drugs and Crime (UNODC). (2012). World Drug Report. Available at https://www.unodc.org/documents/data-and-analysis/WDR2012/WDR_2012_web_small.pdf (accessed August 1, 2016).

U.S. Department of Health and Human Services and U.S. Department of Agriculture. (2015). "2015–2020 Dietary Guidelines for Americans," 8th edition. December 2015. Available at http://health.gov/dietaryguidelines/2015/guidelines (accessed August 1, 2016).

World Health Organization. (2014a). "WHO Global Health Expenditure Atlas." Available at http://www.who.int/health-accounts/atlas2014.pdf (accessed August 1, 2016).

World Health Organization. (2014b). "Global Status Report on Alcohol and Health." Available at http://apps.who.int/iris/bitstream/10665/112736/1/9789240692763_eng.pdf?ua=1 (accessed August 1, 2016).

World Health Organization. (2015). "Alcohol." Available at http://www.who.int/mediacentre/factsheets/fs349/en/ (accessed August 1, 2016).

World Health Organization. (2016a). "HIV/AIDS." Available at http://www.who.int/mediacentre/factsheets/fs360/en/ (accessed August 1, 2016).

World Health Organization. (2016b). "Obesity and Overweight." Available at http://www.who.int/mediacentre/factsheets/fs311/en/ (accessed August 1, 2016).

World Health Organization. (2016c). "Tobacco." Available at http://www.who.int/mediacentre/factsheets/fs339/en/ (accessed August 1, 2016).

World Health Organization Technical Report. (2000). "Obesity: Preventing and Managing the Global Epidemic." Available at whqlibdoc.who.int/trs/WHO_TRS_894.pdf (accessed August 1, 2016).

Yoon, P.W., Bastian, B., Anderson, R.N., Collins, J.L., and Jaffe, H.W. (2014). "Potentially Preventable Deaths from the Five Leading Causes of Death—United States, 2008–2010," *Morbidity and Mortality Weekly Report* 63: 369–392.

Zhou, C. and Zhang, Y. (2012). "The Vast Majority of Medicare Part D Beneficiaries Still Don't Choose the Cheapest Plans that Meet Their Medication Needs," *Health Affairs* 31(10): 2259–2265.

2

A BRIEF OVERVIEW OF BEHAVIORAL ECONOMICS

Thomas Rice, Yaniv Hanoch, and Andrew J. Barnes

The field of behavioral economics has taken policy discussions by storm.[1] Not long ago marginalized in the field of economics, and barely heard of in psychology, it has leaped into prominence, along with terms that are now commonplace in policy discussions: *status quo bias*, *loss aversion*, *defaults*, *choice architecture*, and perhaps most common of all, *nudges*. Early applications were mainly outside of the health care field such as encouraging individuals to save more. Increasingly, however, health services researchers and policymakers have recognized that behavioral economics can help in understanding and optimizing an array of health-related behaviors.

The applications of behavioral economics in Chapters 3 through 13 assume a basic understanding of behavioral economics on the part of the reader. This was intentional, as it helps avoid unnecessary duplication, and more importantly, allows each of the authors to go directly into his or her particular health applications. To ensure readers have a common understanding of the ideas and terminologies, this chapter provides a brief introduction to behavioral economics. After providing a short overview, it presents and discusses several key cognitive biases, and then provides a discussion of how behavior economics tools can be used to improve health decision-making. The chapter ends with a glossary of the terms that were introduced earlier.

Overview of behavioral economics and its antecedents

Behavioral economics draws on criticisms that have been made about the traditional economic model. In that model, individuals are assumed to be rational actors who are able to sift through information (which is in turn assumed to be perfectly and costlessly available) to make best choices in the marketplace that reflect their underlying preferences—that is, they succeed in *maximizing their utility*.

Such a hypothetical person has been called *homo economicus*—rational economic man. Oftentimes, however, to achieve this level of rational decision-making, daily decisions require that these sorts of people "can think like Albert Einstein, store as much memory as IBM's Big Blue, and exercise the willpower of Mahatma Gandhi" (Thaler and Sunstein, 2008, p. 6).

Although one might think that the traditional economic model, based on such strong assumptions, would not stand the test of empirical scrutiny, it nonetheless guided economic thought through almost all of the 20th century. It was, of course, understood that not everyone always sought out, fully understood, and appropriately used relevant available information before making decisions. But deviations were viewed as minor, and as a result, the model was viewed as sufficient for making accurate behavioral predictions. The length to which some economists took this thinking is illustrated in a health-related quotation from Nobel Prize winner in economics Gary Becker and University of Chicago colleague Kevin Murphy: "addictions, even strong ones, are usually rational in the sense of involving forward-looking maximization with stable preferences," and that even though unhappy people often become addicted, "they would be even more unhappy if they were prevented from consuming the addictive goods" (Becker and Murphy, 1988, p. 691). It is hardly surprising that researchers interested in public health have sought an alternative to this type of mindset.

Indeed, there had been detractors over the years. Early on, institutional economists rejected the rational-choice model, with Thorstein Veblen, in 1898, deriding the notion of man as a "lightning calculator of pleasures and pains ..." (Camic and Hodgson, 2011). The work of Herbert Simon (1955) is particularly important. Although trained as a political scientist, Simon was keenly interested in economics (winning a Nobel Prize in the field). He formalized the theory of *bounded rationality*, which recognizes that people do not have the memory and computational wherewithal, much less the time, to use available information to maximize utility. As a result, they instead rely on simpler methods called *heuristics* or rules of thumb. Rather than being maximizers, most people, Simon posited, are instead *satisficers*. As an aside, while one might expect a person who maximizes utility to be happier since she is choosing things that are best rather than "good enough," the research of psychologist Barry Schwartz (2004) suggests the opposite: on average, satisficers not only make better decisions for themselves, but are happier because they experience less stress over the process, and less regret over the options not taken. In any case, Simon would later be followed by other economists and psychologists who developed even more detailed theories for how individuals make decisions that ultimately give rise to observed behaviors.

Behavioral economics and the decision-making process

One of the criticisms of economic theory is that it ignores people's decision-making *process*, focusing instead on observed choices as manifestations of consumers' purported utility maximization. Borrowing from the field of cognitive psychology, behavior economics, in contrast, focuses on how and why people choose what they do, and therefore provides frameworks to those seeking such an alternative to the traditional economic model. Psychologists Daniel Kahneman and Amos Tversky (1979) introduced such an alternative mechanism through their development of Prospect Theory. Among other things, the theory posited that, rather than comparing the utility of two alternatives, people instead focus on the *change* in utility of alternatives relative to a reference point, typically the status quo.

In a subsequent refinement, Tversky and Kahneman argued that people's decisions show both loss aversion (a greater disutility from a loss than they receive in positive utility from a similarly sized gain) as well as diminished sensitivity (where people tend to overweight the utility effects of changes in probability that occur near zero or one, while underweighting those that occur near the middle of the probability distribution) (Tversky and Kahneman, 1992). While too complex to receive full treatment here, the theory can be used to make predictions that are decidedly different than the traditional economic model. In one review, Barberis (2013) provides a number of applications, but interestingly, none from the health care field—further indicating the value of the current book on health care applications.

Another popular model used by behavioral economists is *dual-process theory*. It postulates that people utilize two very different ways of processing information. Kahneman (2011) popularized this idea using the terminology, "System 1" and "System 2." System 1 processing is instantaneous or automatic, like the instinct to jump out of the way of a rapidly approaching car. System 2, in contrast, is deliberative: choosing, say, which model of car to purchase. System 1 decisions, by their nature, cannot follow the traditional economic framework of comparing the benefits and costs of alternatives to maximize utility—although this does not necessarily mean that the choices made are inferior to those made using more deliberative System 2 thinking (Gigerenzer and Goldstein, 1996). It is in System 2 decision-making that traditional economics would anticipate people to succeed in maximizing utility. Due to *cognitive biases*, however, discussed next, this is not always the case.[2]

Cognitive biases

With the advent of new ways of modeling the decision-making process, economists were confronted with the reality that behavioral decisions do not adhere to model predictions for a multitude of reasons. These deviations are often generalized as "cognitive biases." A great deal of effort has gone into researching the many ways in which people's decision-making deviates from that suggested by economic theory. To illustrate, at the time of writing, Wikipedia listed nearly 100 cognitive biases, many of which are quite obscure.[3] Some of the key ones that are helpful in understanding health behaviors are discussed below and in subsequent chapters.

Present bias and salience

It has been suggested that poor health behaviors can arise when people attach more value to things that happen in the present (such as enjoyment of fatty food) and significantly less on those that happen in the future (e.g., increased risk for heart disease). It is natural to put more stock in things that are immediate as opposed to well into the future, and therefore, uncertain. Indeed, traditional economic analysis assigns greater weight to benefits and costs that are closer to the current time period. But people's behavior often shows an extreme *present-bias*. They overemphasize the present, largely because it is more certain and more salient.

Many of the health behaviors discussed in this book—both healthy and unhealthy ones—are affected by present bias. An example of a healthy behavior is exercise. The costs are immediate and often quite salient and therefore typically overweighted. For example, there are short-term physical, psychological, and economic costs (the cost of gym membership; time that could be spent on something else) associated with exercise. Benefits, in contrast, tend to be downstream and underweighted, such as reducing cardiovascular risk and depression (see Chapter 5). In the case of unhealthy behaviors, smoking, substance abuse, unsafe sex, and overeating are all examples in which near term benefits, generally in the realm of pleasure, tend to be overweighted and the associated long-term costs—lung cancer, addiction, sexually-transmitted diseases, obesity—underweighted.

Present bias can be rationalized by *confirmation bias*, in which one pays more attention to evidence that supports one's current views or behaviors. Because it is psychologically distressful to act in a way contrary to one's beliefs, and moreover because it is easier to change the belief than change the behavior, people come up with rationalizations for their behavior (e.g., no one in my family got cancer; what harm could one cigarette do?).[4]

Status quo bias and loss aversion

A key implication of the traditional economic model is that if either the benefits or costs of a good or service change, making an alternative choice more likely to maximize one's utility, then a person will alter her consumption behavior. Indeed, market economics is based on the notion that producers must continually strive to offer the best value or they will lose business to more efficient competitors. Behavioral economics, in contrast, argues that getting people to make changes in their behavior is often far more difficult.

Why might this be the case? One possible explanation is that there may be a good deal of effort involved in making a change: determining the alternatives, obtaining information on the most promising ones, and then actually deciding.[5] The traditional economic model posits that a person should not make a change if the effort involved is not worth it. However, behavioral economics suggests that people have a bias toward the status quo that goes beyond reluctance to incur search costs—indeed, people often fail to make a change when the benefits of doing so exceed the costs.

Prospect Theory offers guidance here. It hypothesizes that people evaluate whether to make a change, using their current position as a reference point. Moreover, the theory postulates that they put much more stock in losses than in gains. As a result, when a person contemplates a change from their current state of affairs, he is likely to evaluate the loss resulting from the change to be greater than the gain.

These cognitive biases may play a role in health-related behaviors. An example from Chapter 9 relates to physicians' inappropriate prescribing of antibiotics. If this is the way they have always practiced medicine, there is a status quo bias. It can be exacerbated by loss aversion—being overly concerned about losing the business of a patient who expects a medication every time he comes to the doctor, and by *action bias*—the feeling that it is better to do something than do nothing. Another example, from Chapter 12, is that people often fail to switch health insurance plans during open enrollment periods even though the plans they possess are inferior to other choices (Sinaiko and Hirth, 2011).

Misestimation of probabilities

While it is natural that people may not be able to accurately estimate probabilities—such as the chance of getting an illness or sustaining an injury—behavioral economics research has found that they often make errors in a predictable direction. One way in which people misestimate probabilities is to overestimate small ones. An example pertains to the side effects of prescription medicines: someone may avoid a useful medication because he overestimates the chances that, in rare instances, it has serious side effects (Berry et al., 2002). Another way in which people misinterpret probabilities is through *anchoring*. This occurs when someone estimates probabilities based on an incorrect initial estimate. Behavioral economics has discovered downright weird manifestations of anchoring. For example, when people were asked to write down the last two digits of their Social Security numbers and then subsequently value a bottle of wine, their purchase bids differed three-fold: those with high two digit numbers valued the wine substantially higher (Ariely et al., 2006). An example in the area of dietary habits relates to package size. Roberto and Kawachi (2014) report that research subjects "used 30% more pasta when they were given a 2-pound box compared to a 1-pound box, and 23% more oil from a 32-ounce bottle than a 16-ounce bottle when making fried chicken" (p. 833), indicating potentially harmful anchoring based on the size of the container in which foods are sold.

Various labeled biases fall under the heading of misestimating probabilities. *Optimism bias* is the unsubstantiated belief that a person will do better than the underlying probabilities. For example, a person in a casino thinking there is a good chance that he'll win the slot machine jackpot, or a smoker believing she'll be less likely to get lung cancer than others. Similarly, *recall bias* occurs when people remember the past incorrectly. Memories, of course, may not be accurate, and this is not a bias in and of itself; one may, for example, overestimate cigarettes consumed one day and underestimate it another. It becomes a bias when a person's recall tends toward a particular direction. Examples would include asking people how often they exercise (where they tend to overestimate), or how often they engage in unprotected sex (where underestimation is common).

Using behavior economics tools to improve decision-making

While the behavioral economics critique of traditional theory is of academic interest, its real appeal is in informing policy. In particular, what does an understanding of cognitive biases tell us about how people can make better decisions?[6] We view the term "policy" in the broadest sense, including not only government but private organizations. For example, much behavioral economics work has been put into assisting companies in getting more of their employees to increase personal savings (Benartzi and Thaler, 2004) or utilizing wellness programs (Volpp et al., 2011).

The policy application of behavioral economics, to some, is encompassed by the notion of the *nudge*. This term was popularized by Richard Thaler and Cass Sunstein (2008) in their book by the same name. The notion comes out of their earlier writings on a philosophical idea they developed called *libertarian paternalism*. Under libertarian paternalism, people are given a full array of alternative choices (the libertarian part) but are guided to ones that are in their best interest (the paternalistic part).[7] Below, we discuss four policy tools that have the potential to improve decision-making. Each could be viewed as a nudge, but it is not necessary to apply that name to any or all of them.

Defaults

Defaults are perhaps the most popular behavioral economics tool, in part because they are often effective in changing behavior substantially. They take advantage of status quo bias or people's natural inertia to stick with a previous decision. The classic example relates to organ donation. In many countries, people's driver's licenses indicate whether they are willing to donate their organs if they are in a fatal car accident. In some places, a person needs to opt in, that is, explicitly choose to be an organ donor. In others, there is an opt-out system: the person is automatically a potential donor unless she explicitly opts out. Whether the default choice is opt in or opt out has a dramatic effect on the proportion of people who are potential donors. Johnson and Goldstein (2004) show that four countries with opt-in systems have organ donation rates ranging from 4 to 28%, while five of the six using opt out have rates of 98% or higher.

A health-related example of defaults comes from the fast food industry. As discussed in Chapter 4, people often stick with an unhealthy food choice because that is what they have always done—for example, having a fast food meal come with a soda or fries rather than choosing alternatives that may be offered such as water and salad. Fast food restaurants can, if they wish, change the default side drinks to the healthier options. McDonald's did that by listing only milk and juice on its menu boards for children's "Happy Meals"; to obtain a soda it is necessary to know if it's available and to ask for it (opt in). This led to a modest reduction in soda consumption over the first year of the program with hopes that this effect will grow over time (Giammona, 2015).

Choice architecture

Defaults are a single example of a broader behavioral economics strategy, known as *choice architecture*. The role of the choice architect is to determine on behalf of others what good decision-making looks like, and then to take the multitude of complex information available and present it more simply, and in a way that facilitates good decision-making in light of both bounded rationality and cognitive biases. This usually involves both reducing the sheer amount of information to a manageable amount, and presenting it in a way that emphasizes what is most important in a format that is easy to rapidly interpret. Another possibility is to *frame* the information in different ways to help foster better understanding. For example researchers have shown that framing information as either gains or losses can promote or inhibit higher vaccination rates. This notion behind choice architecture also stems, in part, from the work of Herbert Simon. Simon believed "a great deal can be learned about rational decision-making … by taking account of the fact that the environment to which it must adapt possess properties that permit further simplification of its choice mechanism" (1956, p. 129).

An example is how information about quality is presented to consumers through health care report cards. Early report cards presented a multitude of complicated metrics but it became quickly apparent that people could not effectively use them. More recently, choice architecture has evolved toward much greater simplicity. The typical way of presenting ratings of health plans and providers is through stars. For example, on Medicare's Hospital Compare website, if one searches for patient ratings of their hospital experience, the first information presented is a simple icon of one through five stars. This is then followed up by more details (e.g., the percent of people who say their doctor's communicated well, percent who say their pain was always well controlled, etc.). Further clicking provides yet even more information, for example, percentages of the above answering "always," "usually," or "sometimes or never."

Sometimes, simply providing *less* information is better. McCormack et al. (2001) conducted an experiment where one group of people received a 100+ page publication called Medicare & You; a second group received even more—the same publication plus a report giving quality scores on all area Medicare managed care plans, and a third group received only a short, abbreviated version of Medicare & You. The control group did not receive any publications. The results were sobering: those receiving more information used it less and were less likely to change health plans. Similarly, as discussed in Chapter 12, good choice is facilitated by a less complex choice environment. In one experiment we tested various ways of simplifying the choice environment, including using symbols rather than dollars to represent premiums and having a choice of three versus nine drug plans. Those facing the more complex information were 23% less likely to choose the least expensive drug plan (Barnes et al., 2012).

Commitments

Traditional economics focuses on monetary incentives—often with success. For example, raising the price of cigarettes and alcohol through heavy sales taxes has, by most accounts, reduced smoking and drinking (Chaloupka et al., 2012; Wagenaar et al., 2010). But some people are not motivated very much by money and even those that are may have large self-control issues. For instance, a person might realize that he has a history of unhealthy behaviors that the best of intentions has not been able to overcome. One way to increase motivation is to put one's reputation at stake. Behavioral economists have taken advantage of this social phenomenon, examining how making *commitments* can influence people to act in ways that they would like.

There are two different types of commitment: public and private. Public ones are likely to be the most effective because a person puts her reputation on the line. Those who make a public commitment and fail to meet it tend to believe that they will look bad in front of others. As described in Chapter 9, doctors who made a public commitment against unnecessary prescribing of antibiotics, by putting up a poster with their picture in their waiting rooms attesting to their commitment to the issue, reduced their antibiotic prescriptions by 20%. An example from a radically different setting is described in Chapter 10 on developing countries. A program for Ethiopian girls ages 10–14, where they and their parents made a public commitment to delay marriage, reduced early marriage by 90% and increased three-fold the likelihood of staying in school.[8]

Private commitments backed up by monetary incentives may also be effective. One prominent example is a website started by academic behavioral economists, *stickk.com*. To use it, a person enters in a goal (e.g., lose 10 pounds in six months) and identifies a sponsor who will monitor her progress. She also enters her credit card number, which will be charged an amount specified in advance, automatically, if she does not meet her goal as adjudicated by a designated sponsor. The website chooses a group of worthy charities that receive the donations when or if the sponsor certifies that the goals are not met. But to increase motivation even more, the person can name an "anti-charity," e.g., the National Rifle Association for those who believe in gun control or the San Francisco Giants if she is a Los Angeles Dodgers fan. At the time of writing the website claimed that people had put more than $25 million on the line, and that it was responsible for 800,000 additional workouts and 15 million fewer cigarettes smoked.

Lotteries

It might be odd to categorize lotteries as behavioral economics interventions. Lotteries are (usually) pecuniary in nature, and it would therefore seem that their use would be a strategy more in line with traditional economics. The reason that they have been embraced by behavioral economists is that in certain circumstances people respond to them more strongly than they would to the equivalent expected-value payment. A 5% chance of winning $100 may be more influential than a flat payment of $5—a surprising result to a traditional economist since most people are risk averse and would be expected to prefer the certain payout.

Indeed, the use of lotteries to incentivize behavior is a popular method of exploiting behavioral economics insights. They rely on leveraging common decision errors that individuals make in dealing with probabilities. Moreover, participation in them is fun. As a result, and discussed in Chapter 7 on HIV, they can be a more cost-effective way of inducing desirable behaviors than conditional cash transfers, which give a fixed reward for obtaining a particular health-related goal. A study from Lesotho found that low-cost lotteries, eligibility for which was based on being tested periodically as HIV-negative, reduced infection rates. Such a strategy can be particularly appealing in low- and middle-income countries, as a small budget can go farther in influencing health behaviors. This dynamic is further accentuated in settings where individuals tend to be risk takers, an example being young people otherwise prone to the "gamble" of unprotected sex, for whom interventions structured around lottery payouts for being tested as HIV-free have been particularly effective (see Chapter 7).

Lotteries can be even more effective if they recognize other factors such as social norms. Loewenstein and colleagues (2013) report on one real-world experiment they conducted. An employer found that it could increase the use of a health assessment tool among its employees by raising the bonus from $25 to $50; doing so increased employee participation from 40 to 44%. The employer was looking for a greater impact, and worked with researchers to design an experiment (which they called a "Dutch Lottery") at another of its worksites. In this type of lottery, winners receive a greater prize if most of their group engages in the desired behavior. Those who completed a health assessment were entered into a $100 lottery, but if 80% or more of their group of 4–8 employees completed the assessment, the lottery stakes were $125. The Dutch Lottery cost the same as the health assessment bonuses, but it increased the use of the assessment tool to 64%.

Limitations of behavioral economics

Over the past two decades or so, behavioral economists have made many contributions that, this book argues, have the potential to improve health behaviors and health. Nevertheless, there are limitations, and the true potential, especially with respect to health applications, is still a subject of speculation. We highlight two concerns here: effectiveness and ethics. Others are discussed in the chapters that follow.

The first limitation is whether behavioral economics interventions have large enough impacts to really make a difference. One area in which they clearly do is with defaults, but as noted below, defaults have their own ethical issues. In contrast, less controversial things like commitment contracts and lotteries, while they can be effective, generally have not shown the ability to change behavior substantially. Many behavioral economists recognize this shortcoming, noting that such interventions are not meant to replace traditional economic and regulatory tools. Adam Oliver, for example, concluded that behavioral economics,

> can perhaps best be thought of as offering a library of tools, not all of which can be used at any specific time, but each of which may be of use in some particular contexts. Behavioural economics is not a panacea, but by using the insights from human psychology that are embedded in the approach, academics and policy makers may be able to design interventions that—in some circumstances—are relatively well equipped to motivate people to behave in ways that are better for themselves, and for society at large.
>
> *(Oliver, 2012)*

The ethical issues are of perhaps greater concern. Defaults are an excellent example. We saw earlier that, in countries where being an organ donor is the default, almost everyone becomes a donor. But this does not necessarily reflect their preferences. Rather, it may simply reflect status quo bias or a lack of understanding of how to opt out of donations. By making it the default, government may be ascribing preferences to people that they do not actually have.

There are other aspects of choice architecture that also raise ethical issues. One of us has suggested elsewhere that an effective way to improve people's decision-making is to reduce the number of health insurance plans available (Rice and Cummings, 2010). If the "bad" choices in the eyes of the choice architect are weeded out before the open enrollment period, then people should be more likely to choose a plan that, in the view of the choice architect, would raise their utility. This will mean, however, that choices that some people would have picked are not available. Related to this is the fact that the choice architects who make these decisions on behalf of individuals are often regulators and thus not subject to any democratic accountability (Zimmerman, 2015).

Conclusion

The upsurge of interest in behavioral economics is an exciting development in the field of health behavior. While the use of traditional economic and regulatory tools can result in greater adoption of health-promoting behaviors, they are often limited because people act in ways inconsistent with what those models predicted. Behavioral economics leverages insights from the fields of psychology in an attempt to better understand how people actually behave in the real world. Moreover, its insights can be used to develop policies to help improve people's behavior in many aspects of life, including health.

There is, as yet, little agreement on exactly how the field is defined, nor is there an overarching, agreed-upon theoretical basis for it.[9] What does appear to be true, however, is that the field of economics now recognizes that the cognitive biases discussed in this chapter are potentially important phenomena that deserve attention by scholars. In turn, this means that policies that go beyond those traditionally recommended by economists—mainly, interventions aimed at altering price or enhancing available information—have a place in serious policy discussions. Armed with a bigger and better toolbox, policymakers are using behavioral economics to find new ways to make healthy behaviors easier for consumers to adopt. The following chapters document such contributions in almost a dozen different aspects of health care.

Glossary

Note: Most terms in italics are defined elsewhere in the Glossary.

Action bias: A cognitive bias in which people favor action over inaction, even if it is not in their best interest. An example is a patient insisting that a physician order a test even when it is not medically justified.

Anchoring: A behavioral phenomenon in which, for reasons unrelated to the problem at hand, certain numbers are cognitively salient, which affects one's beliefs or decisions, typically in an inappropriate matter. Because these anchors are unrelated to the benefits and costs of the decisions in hand, their salience leads to mistakes. An example from the text provides an illustration: when subjects were asked to write down the last two digits of their Social Security numbers first, those with high numbers bid far higher for bottles of wine, textbooks, or computer trackballs (Ariely et al., 2006).

Bounded rationality: A theory, originally developed by Herbert Simon, which posits that people's decisions do not conform to the traditional economic theory of utility maximization because they are subject to cognitive limits in their ability to process available information, and because choice environments are not structured in a way that facilitates tractable decision-making. People instead use *heuristics* to pare down available information, and *satisfice* rather than maximize.

Choice architecture: The choice architect *frames* a choice and chooses and summarizes information that is ultimately presented to consumers to help them make particular decisions—taking into account people's *cognitive biases* and recognizing that they exhibit *bounded rationality.* In one example, a choice architect might winnow down the amount of information and highlight it in a way to facilitate comparisons of alternative choices.

Cognitive biases: Systematic deviations from what traditional economic theory defines as rational behavior. These biases can result in people's behavior being inconsistent with the predictions of the traditional theory. Behavioral economics posits that recognition of these biases has important implications regarding which types of public policies should be considered. The cognitive biases that appear to play the greatest roles in health behaviors, and which therefore receive the most attention in this book, are *present bias* and *status quo bias*.

Confirmation bias: A cognitive bias where people put more stock in information that is concordant with their pre-existing beliefs or behaviors. In doing so, they are likely to discount contrary evidence and continue with their current behavior, resulting in *status quo bias*.

Defaults: An alternative that is assigned to a person who does not actively make a choice. *Choice architects* may set a default that they believe is in people's best interests, taking advantage of *status quo bias*—the idea being that most people will go along with the default that is assigned. In most cases, people are allowed to opt out of the default. One classic example is the (legally enforceable) assumption that people are willing to serve as organ donors in case of a fatal accident, unless they explicitly state otherwise.

Dual-process theory: A psychological *heuristic* positing that people process information in two different ways. Although there are several different versions of dual-process theory, one version, put forward by Daniel Kahneman, characterizes one type of mental process as being instinctual or automatic (System 1) and the other as being more deliberative (System 2). The assumption that people use classical maximization techniques (e.g., sifting through information related to every alternative) does not hold for System 1 decisions. Rather, they rely on what they already know or do and, at best, use *heuristics*.

Framing: A frame is a cognitive apparatus, or cognitive context, that helps people make sense of new information. Any decision problem, including a choice, is framed by both the information and the other choices that surround it. This surrounding information can influence a choice by making one option look better by comparison to other choices or more attractive because of associations with particular kinds of information. Behavioral economics experiments have shown, for example, that students are more likely to take fruit and vegetables when they come at the beginning of the lunch line than the end.

Heuristics: A simple model of how the world works or an approximate rule of thumb for decision-making. Rather than considering all options and information about them, people often use heuristics. This is particularly true for System 1 thinking (see *dual-process theory*, above). There is considerable debate as to whether the use of heuristics leads to better or worse decision-making. On the one hand, mental shortcuts can accentuate reliance on cognitive biases (e.g., *status quo bias, present bias*). Alternatively, given people's limited mental capacity and time, the use of heuristics may be necessary in making good decisions.

Homo economicus: Literally, "rational economic man." According to traditional economic theory, people have perfect information, which they use to maximize their utility, defined by fixed and immutable preferences. Homo economicus is free of the *cognitive biases* that the field of behavioral economics studies.

Libertarian paternalism: A theory of public policy proposed by Richard Thaler and Cass Sunstein that seeks to balance social welfare and individual liberty. Because of the many challenges facing consumers (e.g., complex information, cognitive biases), they might make poor decisions. Libertarian paternalism suggests that people continue to enjoy all available choices, but that they be guided toward those that, in the view of the policymaker, are most in that person's (or society's) interest. The policymaker can be from the private sector (e.g., the employee benefits manager of a company's pension plan) or the public sector.

Loss aversion: When much greater disutility is associated with a loss than the utility associated with a similarly sized gain. This can result in *status quo bias* because people are overly concerned they will lose something by making a change—overshadowing the potential good that will be derived from it.

Nudge: A term used by Richard Thaler and Cass Sunstein, from their book of the same name. They suggest that policymakers "nudge" people toward better decisions. Based on their theory of *libertarian paternalism*, people would retain all choice options but should be nudged toward those that are in their (or society's) best interest. A common form of nudging is to assign people the *default* choice that the policymakers view to be in their best interest, requiring them to actively opt out if they want a different option. Nudging and *libertarian paternalism* are controversial, in part because people could be nudged toward things that the policymaker wants rather than those things that are in the consumer's best interest.

Optimism bias: Unwarranted belief that one will fare better than the underlying odds, or better than others will do, which can in turn cloud effective decision-making. An example is a drunk driver who believes that his chance of making a mistake while driving is significantly lower than the actual probabilities, or than the chances of others.

Present bias: An undue preference for the present, brought about largely by the *salience* of events that are close at hand compared to those far in the future. With regard to health behaviors, present bias can result in overweighting the utility of pleasures such as eating fatty foods or using narcotics, while underweighting the consequences of obesity and addiction because they are manifested later.

Prospect theory: A theory, originally developed by Daniel Kahneman and Amos Tversky, which offers predictions about behavior that differ from those derived from the traditional economic theory of utility maximization.

Among other things, the theory suggests that people make decisions not based on their perception of the absolute advantages and disadvantages of competing alternatives, but rather on their perception of how such competing alternatives *change* their well-being, relative to a reference point, which is usually the status quo. Moreover, it suggests that people put much more stock in losses than in gains (*loss aversion*) and overweight the probability of rare events occurring (and underweight common ones).

Recall bias: The tendency to remember past events or habits not only incorrectly, but with a tendency in a particular direction. Much health information is based on self-assessments from surveys. Asking people, for example, how much they smoked or ate fatty foods in the past often leads to underestimates of the actual behavior because people's recollections are biased toward remembering when they behaved well.

Salience: Any thought that is, in a given situation, more readily available to the brain than it would otherwise be is said to be salient. For example, if you ask someone to say "silk" five times and then ask them what a cow drinks, they may say "milk" because the sound "-ilk" is salient. In the context of behavioral economics, certain benefits or costs may be especially apparent, or salient, to the decision maker. These tend to be those that are current or immediately in the future. Things that are more salient tend often to be given much greater weight, resulting in *present bias*. Salience can be enhanced by *priming*.

Satisficing: The traditional economic theory of choice is that consumers always seek to maximize their utility, which would require considering all alternatives available. Satisficing in contrast, involves examining options in order of their *salience*, and choosing the first option that is good enough. Once a good-enough option has been identified then further, typically less *salient*, options are not explored.

Status quo bias: People's tendency to favor what they already have. Such a tendency could be consistent with traditional economic theory: there is less uncertainty because a person knows more about the things already possessed, and there may be costs in terms of time or effort in making a change. The bias, then, refers to people being *overly* wedded to what they already have or chose in the past, such that they miss opportunities that would make them better off.

Notes

1 A Google Scholar search of both the American and British spellings of the term came up with about 5,200 articles per month mentioning the term in the 2010s, a 25-fold increase compared to the 1990s.

2 In discussing the dual-process theory, Kahneman admits that we have just a single brain, and that the System 1–System 2 model is simply a heuristic. The theory has been criticized by others as raising both logical and ethical problems (Zimmerman, 2015).

3 One of the authors' favorites is the "IKEA effect": The tendency for people to place a disproportionately high value on objects that they partially assembled themselves, such as furniture from IKEA, regardless of the quality of the end result (https://en.wikipedia.org/wiki/List_of_cognitive_biases).

4 This type of thinking is known as *cognitive dissonance*.

5 With regard to the latter, behavioral economists have also studied *decision fatigue*, the idea being that having to make decision after decision can be so mentally enervating that the quality of the decisions made falls over time (Danziger et al., 2011).

6 Also of importance is the issue of "better decisions" from whose perspective. Normally behavioral economists are thinking of the person—that is, to help maximize utility. But behavioral economics tools are also sometimes used to encourage behavior that, some feel, would be in society's best interest. Increasing people's tendency to recycle is an example. To the extent that individual and social welfare are at conflict, behavioral economics interventions can be both controversial and raise ethical issues, as discussed at the end of the chapter.

7 Gigerenzer (2015) objects to the conflation of libertarian paternalism and nudging. While he does not object to the notion of nudges, he argues against the libertarian paternalistic view that people need to be nudged because they are irrational. Rather, Gigerenzer argues, it "focuses the blame for societal problems exclusively on the individual mind, closing our eyes to institutions that steer individual behavior so that they can take advantage of it, and it misleadingly suggests that a more sustainable solution, educating people, is a hopeless endeavor" (p. 363).

8 It should be noted that the families of the girls who completed the program also received a goat "to help offset the costs of delaying marriage" which in turn also could have been partly responsible for the results.

9 Prospect Theory is sometimes cited as a theoretical basis for behavioral economics. By one account, Kahneman and Tversky's (1979) original article on the concept is the fifth most referenced article in the history of social sciences, with over 41,000 citations on Google Scholar as of June 2016. See http://blogs.lse.ac.uk/impactofsocialsciences/2016/05/12/what-are-the-most-cited-publications-in-the-social-sciences-according-to-google-scholar/.

References

Ariely, D., Loewenstein, G., and Prelec, D. (2006). "Tom Sawyer and the Construction of Value," *Journal of Economic Behavior & Organization* 60: 1–10.

Barberis, N.C. (2013). "Thirty Years of Prospect Theory in Economics: A Review and Assessment," *Journal of Economic Perspectives* 27(1): 173–196.

Barnes, A., Hanoch, Y., Wood, S., Liu, P.-J., and Rice, T. (2012). "One Fish, Two Fish, Red Fish, Blue Fish: Effects of Price Frames, Brand Names, and Choice Set Size in Medicare Part D Insurance Plan Decisions," *Medical Care Research and Review* 69(4): 460–473.

Becker, G.S. and Murphy, K.M. (1988). "A Theory of Rational Addiction," *Journal of Political Economy* 96(4): 675–700.

Benartzi, S. and Thaler, R. (2004). "Save More Tomorrow: Using Behavioral Economics to Increase Employee Saving," *Journal of Political Economy* 112(Suppl.): S164–S187.

Berry, D.C., Knapp, P., and Raynor, D.K. (2002). "Provision of Information about Drug Side-Effects to Patients," *Lancet* 359: 853–854.

Camic, C. and Hodgson, G.M. (eds) (2011). *Essential Writings of Thorstein Veblen*. New York: Routledge.

Chaloupka, F.J., Yurekli, A., and Fong, G.T. (2012). "Tobacco Taxes as a Tobacco Control Strategy," *Tobacco Control* 21: 172–180.

Danziger, S., Levav, J., and Avnaim-Pesso, L. (2011). "Extraneous Factors in Judicial Decisions, *PNAS* 108: 6889–6892.

Giammona, C. (2015). "McDonald's Finds Many Kids Still Want Soda with Happy Meals." Available at http://www.bloomberg.com/news/articles/2015-06-25/mcdonald-s-finds-many-kids-still-want-soda-with-happy-meals (accessed December 31, 2016).

Gigerenzer, G. (2015). "On the Supposed Evidence for Libertarian Paternalism," *Review of Philosophy and Psychology* 6(3): 361–383.

Gigerenzer, G. and Goldstein, D.G. (1996). "Reasoning the Fast and Frugal Way: Models of Bounded Rationality," *Psychological Review* 103(4): 650–669.

Johnson, E.J. and Goldstein, D.G. (2004). "Defaults and Donation Decisions," *Transplantation* 78(12): 1713–1716.

Kahneman, D. (2011). *Thinking, Fast and Slow.* New York: Farrar, Strauss and Giroux.

Kahneman, D. and Tversky, A. (1979). "Prospect Theory: An Analysis of Decision Under Risk," *Econometrica* 47: 263–292.

Loewenstein G,. Asch, D.A., and Volpp, K.G. (2013). "Behavioral Economics Holds Potential to Deliver Better Results for Patients, Insurers, and Employers," *Health Affairs* 32(7): 1244–1250.

McCormack, L.A., Garfinkel, S.A., Hibbard, J.H., Norton, E.C., and Bayen, U.J. (2001). "Health Plan Decision Making with New Medicare Information Materials," *Health Services Research* 36 (3): 531–554.

Oliver, A. (2012). "Whither Behavioural Economic Policy," London School of Economics and Political Science. Available at http://blogs.lse.ac.uk/healthandsocialcare/2012/04/12/whither-behavioural-economic-policy/ (accessed December 31, 2016).

Rice, T. and Cummings, J. (2010). "Reducing the Number of Drug Plans for Seniors: A Proposal and Analysis of Three Case Studies," *Journal of Health Politics, Policy and Law* 35(6): pp. 961–997.

Roberto, C.A. and Kawachi, I. (2014). "Use of Psychology and Behavioral Economics to Promote Healthy Eating," *American Journal of Preventive Medicine* 47(6): 832–837.

Schwartz, B. (2004). *The Paradox of Choice: Why More Is Less.* New York: HarperCollins.

Simon, H.A. (1955). "A Behavioral Model of Rational Choice," *Quarterly Journal of Economics* 69: 99–118.

Simon, H.A. (1956). "Rational Choice and the Structure of the Environment," *Psychological Review* 63: 129–138.

Sinaiko, A.D. and Hirth, R.A. (2011). "Consumers, Health Insurance and Dominated Choices," *Journal of Health Economics*, 30(2): 450–457.

Thaler, R.H. and Sunstein, C.R. (2008). *Nudge: Improving Decisions About Health, Wealth, and Happiness.* New Haven, CT: Yale University Press.

Tversky, A. and Kahneman, D. (1992). "Advances in Prospect Theory: Cumulative Representation of Uncertainty," *Journal of Risk and Uncertainty* 5(40): 297–323.

Volpp, K.G., Asch, D.A., Galvin, R., and Loewenstein, G. (2011). "Redesigning Employee Health Incentives—Lessons from Behavioral Economics," *New England Journal of Medicine* 365: 388–390.

Wagenaar, A.C., Tobler, A.L., and Komro, K.A. (2010). "Effect of Alcohol Tax and Price Policies on Morbidity and Mortality: A Systematic Review," *American Journal of Public Health* 100(11): 2270–2278.

Zimmerman, F.J. (2015). "From Choice Architecture to Policy Infrastructure: Multilevel Theory and the Political Economy of Health Behaviors," in I. Kawachi and C. Roberto (eds), *Behavioral Economics and Public Health.* Oxford, UK: Oxford University Press: 299–341.

PART II
Shaping health behaviors

3

THE BEHAVIORAL ECONOMICS OF TOBACCO PRODUCTS

Innovations in laboratory methods to inform regulatory science

Warren K. Bickel, Lara N. Moody, Sarah E. Snider, Alexandra M. Mellis, Jeffrey S. Stein, and Amanda J. Quisenberry

Introduction

Although the smoking of combustible tobacco products remains a leading cause of preventable mortality and morbidity, tobacco control is entering a period of disruptive innovation resulting from two changes that may fundamentally alter the tobacco marketplace. First, the Food and Drug Administration (FDA) was granted authority in 2009 to regulate tobacco products to benefit public health. With this authority, the FDA can regulate the amount of nicotine in tobacco products as well as a variety of other design features (e.g., flavors). Second, an increasing array of new products such as electronic cigarettes (e-cigarettes) has led to an ever-more complex tobacco marketplace. Consumption of these new products may result in less health hazard than conventional cigarettes, although more data is necessary to clearly demonstrate this.

For the FDA to effectively regulate tobacco products and improve public health, a great deal of tobacco regulatory science will be necessary. Among the questions that must be answered, in addition to the health consequences of these new products, is the likelihood that they will be abused; that is, will a current smoker find these products sufficiently reinforcing to promote their use? Another important question is to what extent do these new products lead to substitution; that is, to what extent do these products share the qualities important to current smokers such that smokers will be likely to switch?

Tobacco control benefiting the public health could be advanced by novel methods that permit estimates of the abuse liability of new products by current smokers and how these products may shift consumption (e.g., substitution). Behavioral economics provides methods to understand the abuse liability and the interaction among tobacco and other products. Over the last 25 years, these methods have evolved substantially over four major epochs of innovation: (1) operant self-administration experiments, (2) the hypothetical purchase task, (3) naturalistic demand assessments, and (4) the Experimental Tobacco Marketplace. In this chapter, after reviewing key concepts of behavioral economics, we will illustrate each methodological advance and how it has contributed to the understanding of tobacco consumption (Bickel and Madden, 1998).

Why behavioral economics models?

The utility of behavioral economics models derives from the ability to look at a more extensive range of conditions than is typically available in standard economics studies. Specifically, standard economics typically retrospectively examines a limited range of prices that have occurred in the natural economy. Therefore, standard economics analyses can only examine historical phenomena and cannot prospectively contribute to tobacco regulatory science. In contrast, behavioral economics laboratory studies permit experimental examination of a broad range of prices in the presence of alternative products. These methods provide exquisite control, permitting detailed analytical understanding of a diverse set of economic and environmental conditions on tobacco consumption and, consequently, a prospective analysis of key determinants of tobacco regulatory science.

The behavioral economics of abuse liability

Throughout each of the four epochs of innovation, behavioral economics methods have provided assessments of the abuse liability of existing and emerging tobacco products. Defined broadly, abuse liability describes the degree to which a substance is likely to produce persistent, problematic patterns of use (Food and Drug Administration, 2010). More specifically, abuse liability is a multivariate construct drawing from decades of research in the natural, physical, and social sciences, where the degree of abuse liability depends on a number of factors. These factors, among others, include a tobacco product's subjective effects, speed of onset of those effects, the product's reinforcing value, and the likelihood it would produce physical or psychological dependence (Balster, 1991; Carter et al., 2009; Henningfield and Keenan, 1993; Stitzer and de Wit, 1998). Additional product constituents may also act synergistically with nicotine to increase abuse liability beyond that of nicotine alone (e.g., acetaldehyde and monoamine oxidase inhibitors in tobacco smoke; Guillem et al., 2005; Talhout et al., 2007). Moreover, the potential influence of environmental factors in determining abuse liability, such as social acceptance, product marketing, and ease of purchase, has also been recognized (Carter and Griffiths, 2009; O'Brien, 1996; World Health Organization, 2008).

Each component of abuse liability may be assessed using distinct methodology (Carter et al., 2009; Carter and Griffiths, 2009). For example, pharmaco-kinetic and -dynamic factors, such as the magnitude and speed of onset of drug effects, may be tested using biological sampling and self-report measures; the likelihood of physical dependence may be examined through the assessment of withdrawal following acute or chronic drug administration; and reinforcing value may be examined through simulated or real drug self-administration procedures. Assessment of self-administration provides the broadest understanding of abuse liability, as decisions to self-administer a drug (assuming sufficient prior experience) are influenced by all additional factors in the construct (e.g., speed of onset and magnitude of effects).

Own-price elasticity of demand

Drug self-administration can be uniquely understood with behavioral economics. The basis of this unique understanding is through assessment of a drug's *own-price elasticity* of demand, or sensitivity of consumption of a commodity (e.g., cigarettes) to increases in unit price. Unit price, a cost–benefit ratio, is broadly defined as the behavioral or monetary cost required to obtain one unit of a commodity (DeGrandpre, Bickel, Hughes et al., 1993; Hursh, 1980). Thus, the degree to which consumption of a given commodity is sensitive to changes in its own price (i.e., own-price elasticity) provides an index of that commodity's value, with lower levels of elasticity representing a greater willingness to defend consumption of that commodity against increasing price (Hursh and Roma, 2013; Hursh and Silberberg, 2008; Hursh and Winger, 1995). When applied to drugs of abuse, measures of elasticity reflect one component of abuse liability (Carter et al., 2009).

Own-price elasticity of demand is typically quantified using methods of least squares regression. Historically, linear demand models have been used to estimate demand elasticity in the behavioral economics laboratory (see Hursh et al., 1989 for a description of the linear model). However, an exponential model of demand is now more frequently used, which describes the nonlinear relationship between consumption and price evident at the individual-subject level (Hursh and Silberberg, 2008):

$$log_{10}Q = log_{10}Q_0 + k(e^{-\alpha Q_0 P} - 1), \qquad (1)$$

where Q is consumption, P is price, k is span of consumption in $\log_{10}$ units, and Q_0 and α are free parameters describing demand intensity (i.e., consumption at minimal price) and demand elasticity, respectively. Here, we note that the Q_0 and α parameters may be interpreted independently of one another.

Although linear and nonlinear equations are useful in quantifying demand elasticity, we have recently proposed and validated a modified version of Equation 1 that remediates problems posed when participants contribute zero values in consumption (e.g., at the highest prices). Because zero values cannot be log transformed, application of Equation 1 to demand data requires either elimination of zeroes or replacement of zeroes with nominal values (e.g., 0.01); both of these strategies, however, can drastically affect demand estimates depending on the number of obtained zero values or the precise nominal values used (see Koffarnus, Franck, et al., 2015). Use of the following modified equation, however, eliminates these problems (Koffarnus, Franck, et al., 2015):

$$Q = Q_0 * 10^{k(e^{-\alpha Q_0 P} - 1)}, \tag{2}$$

in which all parameters are otherwise identical to those described for Equation 1; however, the equation appears as an exponent of 10, removing the need for log transformation and therefore allowing analysis of unaltered consumption values, including zeros.

An additional analytical challenge comes when participants occasionally contribute consumption data that are not systematically affected by price (i.e., either invariant, highly variable, or no consumption across prices). In some cases, such data may be an important property of the commodity or population under investigation; however, in others, such data are likely produced by extraneous variability not germane to research aims (e.g., inattention or failure to understand task instructions) and may obscure effects of the variable(s) under investigation. We recently proposed and validated a method for identifying such data, assuming only that consumption of a commodity should decrease monotonically with increasing price (Stein et al., 2015). Use of this method, as well as related guidelines for handling nonsystematic data, may serve to enhance accuracy of demand estimates and therefore improve the ability to detect effects of clinical treatments and other variables that influence demand for tobacco products.

Cross-price elasticity of demand

In contrast to own-price elasticity, measures of *cross-price elasticity* describe the interaction between the reductions in consumption of one price-manipulated product (e.g., cigarettes) and the consumption of one or more constant-priced alternative products (e.g., e-cigarettes). The economic relationship between these products may be categorized along a continuum in which the constant-priced alternative products serve as either: (1) substitutes, at one end of the continuum (i.e., consumption of a constant-priced alternative product increases to compensate for reduction in consumption of a price-manipulated product), (2) independents (i.e., consumption of the constant-priced alternative product does not change), or (3) complements, at the other end of the continuum (i.e., consumption of both the constant-priced alternative product *and* the price-manipulated product decrease).

Like own-price elasticity, cross-price elasticity of demand has been historically quantified using linear models (Allison, 1983; DeGrandpre et al., 1994); however, reflecting its nonlinear form, cross-price elasticity may be quantified by fitting consumption curves for constant-priced alternative commodities using the exponential equation (Hursh and Roma, 2013, 2015):

$$Q = log_{10}(Q_{alone}) + Ie^{-\beta P}, \tag{3}$$

in which the free parameter Q_{alone} is asymptotic consumption of a constant-priced, alternative commodity when the price (P) of a price-manipulated commodity is infinitely high; I is the y-axis span of consumption in $\log_{10}$ units, describing the degree of interaction between commodities; and β is sensitivity of alternative product consumption to reductions in consumption of the price-manipulated commodity. Here, negative values of the I term indicate substitution between commodities and positive values indicate complementarity.

In the following sections, we review historical and contemporary methods used to assess the behavioral economics demand measures outlined above. We organize these sections according to the four epochs of methods innovation, highlighting important findings from each epoch that have contributed to our understanding of tobacco regulatory science.

Behavioral economics 1.0: operant self-administration

The first behavioral economics methods within the human laboratory were modeled after preclinical operant procedures, such as pigeons pecking a key to obtain food reinforcement (Hursh, 1984). These procedures were used to examine the effects of price on tobacco self-administration in the clinical laboratory. Importantly, researchers obtained reliable and parsimonious results, even while utilizing different procedural details, and developed predictions of how these factors may be generalized to the real world (Bickel et al., 1993). Moreover, participants' consumption of cigarettes is reliably sensitive to increases in price (DeGrandpre et al., 1992) and consistent with the broader concept of consumer demand. In the laboratory, unit price, or the amount of responses necessary to receive the reinforcer, may be increased using one of two methods: (1) utilizing progressive ratio schedules (Stafford et al., 1998) in which the participant is required to pull a plunger a set amount of times in order to receive a puff of a cigarette; or (2) decreasing the magnitude of the reinforcer (i.e., minimizing puff volume) while keeping the fixed ratio requirement (i.e., plunger pulls) constant. When these two methods of manipulating unit price were directly compared, cigarette puff consumption was consistent, regardless of whether the ratio requirement was increased or puff magnitude was decreased (Bickel et al., 1991). Interestingly, participants also consumed consistent numbers of puffs at each unit price, regardless of incremental progressive ratio increases or random ratio dictated unit price (Giordano et al., 2001). Interventions attempting to manipulate puff consumption were largely unsuccessful at the time, indicating that puff consumption is consistently sensitive to unit price. However, puff intake was manipulated under specific conditions. For example, cigarette puff intake was more elastic (i.e., decreased more readily with increasing unit prices) when smokers were acutely deprived (Madden and Bickel, 1999). Although these effects were small and waned over time, they supported the notion that consumption may be manipulated temporarily depending upon the circumstances of the consumers' environment. Finally, while different commodities such as conventional cigarettes and de-nicotinized cigarettes are sensitive to increasing prices, interactions between commodities can occur when alternative rewards, such as money, are available concurrently (Shahan et al., 2001).

Preference for alternatives

In a review and re-analysis of 16 studies, findings from the drug self-administration laboratory using different species, drug classes, and procedures demonstrated that concurrently available commodities could act as substitutes, independents, or complements (Bickel et al., 1995). The interaction type between commodities occurs depending on the concurrently available alternatives, their magnitude, and the relative price of each (Bickel et al., 1995). For example, when conventional and de-nicotinized cigarettes were available alone with increasing prices, each produced similar elasticities of demand, consistent with price sensitivity. However, when participants were given the choice between the two concurrently available commodities at equivalent prices (i.e., fixed ratio requirements), participants chose conventional cigarettes significantly more over de-nicotinized cigarettes (Shahan et al., 1999), indicative of greater valuation of conventional cigarettes. Interestingly, in other cases the presence of some commodities may have no effect on consumption at all, in which case the commodities are considered independents. For example, when cigarette puffs and money were available concurrently, the increase in unit price of cigarette puffs did little to affect preferences for the money condition (DeGrandpre et al., 1994). Similar operant procedures demonstrated that when unit price for cigarette puffs (i.e., fixed ratio) was increased, coffee consumption decreased alongside cigarette consumption, indicative of complementary commodities (Bickel et al., 1992).

Increasingly complex combination studies were introduced to the laboratory to answer questions such as, "Do smokers differentially consume tobacco products when alternatives are available?" For example, cigarettes, nicotine gum, and/or money were presented alone first and then in combination (Johnson and Bickel, 2003). As illustrated in Figure 3.1, cigarette puff consumption decreased with increasing unit price in all conditions, as expected. Interestingly, when cigarettes and gum were presented concurrently, gum substituted for cigarettes as the price of cigarettes increased, whereas money was again independent of the cigarette price increase. When all three commodities were available together, the total consumption of cigarette puffs decreased, supporting the notion that the presence of substitute and independent reinforcers can reduce cigarette intake (Johnson and Bickel, 2003). An expansion of these methods found that with increasing prices of conventional cigarettes, both de-nicotinized cigarettes and nicotine gum functioned as substitutes (i.e., when only two commodities were available). However when all three commodities were available together, de-nicotinized cigarettes were more reliably chosen over the gum as an alternative (Johnson, Bickel, and Kirshenbaum, 2004). Again, these combination studies highlight how consumers will alter their purchasing when presented with alternatives and can help researchers predict consumer behavior in more realistic environments.

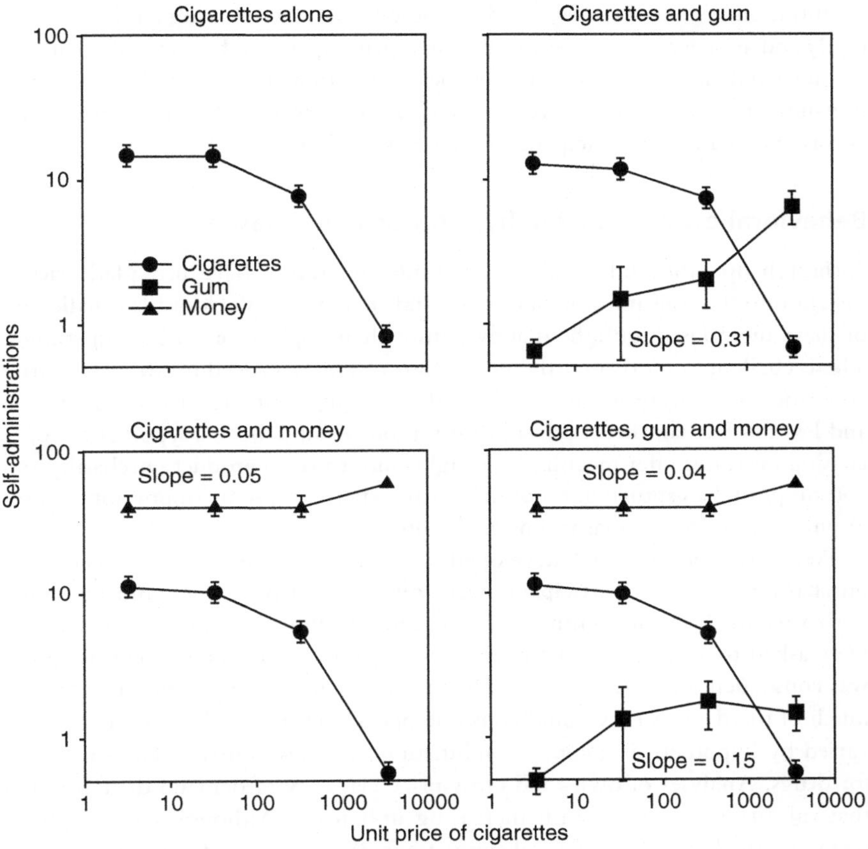

FIGURE 3.1 Group mean (±SEM; n=11) self-administrations of cigarette puffs, nicotine gum, and/or money presented as a function of increasing unit price for cigarettes. Each of four phases demonstrates differential self-administration dependent on the commodities available. Consistent with early work using linear demand analysis with the log of both axes, zero values were replaced with trivially small nonzero values (in this case, 0.5).

Although operant procedures such as those described above allow researchers to generate demand curves and directly measure the interactions between concurrently available commodities and their self-administration in the laboratory, weaknesses exist for this approach. First, multi-session laboratory-based progressive ratio procedures are highly staff- and time-intensive, each session lasting several hours, and requiring expensive specialized equipment. Second, in single session studies, breakpoints, where individuals stop responding for the reinforcer, can be obtained, however demand curves cannot be generated and participants may become fatigued and/or satiated (Hursh and Silberberg, 2008). Practical barriers, such as the cost of building or renovating laboratory space with air filtration systems to safely conduct these laboratory smoking studies, can be costly and are necessary to reduce the risk of participants and study staff breathing second-hand smoke. While some of these limitations remain in the preclinical non-human laboratories, more recent clinical work has obtained reliable and consistent data using participant self-report measures.

Behavioral economics 2.0: hypothetical purchase task

Although operant studies of drug self-administration have provided critical insight into the consumption of tobacco and alternatives, the logistical challenges of these methods (highlighted above) limit their application in human studies. These challenges motivated the generation of a new procedure to model drug consumption. This procedure is a hypothetical questionnaire that is less invasive and less costly than direct self-administration of substances. Hypothetical purchasing measures offer meaningful insights into tobacco product purchasing and consumption by capturing, if not the entire experience, some important features of purchasing decisions (as in Epstein, 1986).

Petry and Bickel (1998) developed a novel simulation task to model drug purchasing decisions. Participants were given imitation money and presented with various drugs at commonly used amounts and street prices. Participants were asked to assume that the drug was their only source and its consumption was consequence free (e.g., legal). They were also instructed to purchase only as much of the drug as they could consume personally in a 24-hour period. Trials varied by hypothetical income, combination of drugs available, and unit price for drugs. Analyses of own- and cross-price elasticity of demand demonstrated that substitutes emerged with increasing unit price. Although these findings were obtained from hypothetical simulations, they concurred with other observations of drug consumption. These results suggested the hypothetical purchase questionnaire captured important features of purchasing behavior and set the occasion for the later application to studies of tobacco consumption.

A second study of 16 opiate-dependent cigarette smokers, enrolled in outpatient buprenorphine therapy, evaluated the relative rate of reinforcement between heroin and tobacco products (Jacobs and Bickel, 1999). This model used a purely questionnaire-based method, and participants completed paper questionnaires where heroin and cigarettes were available at 15 price points. Traditional analyses of overall reinforcement rate and overall purchasing indicated that heroin was a more effective reinforcer than cigarettes. However, preference between heroin and cigarettes varied over prices in concurrent conditions. The relative location of the two demand curves when heroin and tobacco were singly available predicted preference when they are concurrently available. Thus, the hypothetical purchase task offered an efficient method to reveal insights into the interplay between cigarette and other drug consumption, and a rapid evaluation of tobacco purchasing.

MacKillop et al. (2008) investigated cigarette purchasing in college smokers using a hypothetical cigarette purchase task at 19 prices ranging from $0 (free) to $1,120 per cigarette. Overall cigarette purchasing declined as the unit price of cigarettes increased, with demand measures corresponding with nicotine dependence scores and self-reported daily smoking. When asking participants to purchase using different allotted incomes, MacKillop et al. did not see significant correlation between income and cigarette purchasing behavior, indicating that income did not affect the value of cigarettes. Critically, observations of tobacco purchasing from this task corresponded to most indices of real-world tobacco consumption and dependence, with overall expenditure on cigarettes significantly differing as a function of nicotine dependence, as illustrated in Figure 3.2.

Modern adaptations of the hypothetical purchase task have also been used to offer insights into demand for alternative tobacco products among cigarette smokers (O'Connor et al., 2014), and also the potential substitution of e-cigarettes for combustible cigarettes (Grace et al., 2015). Although the hypothetical purchase task may involve a more rapid, electronic assessment (Wilson et al., 2015) rather than the paper-and-pencil task first developed, it is still limited by its hypothetical nature. Individuals' responses to simulated economic decisions may not correspond to their day-to-day choices in the broader marketplace. The

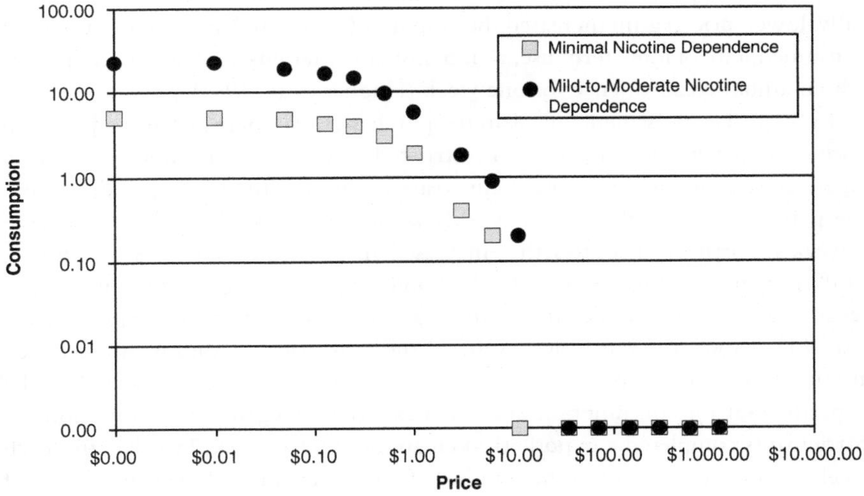

FIGURE 3.2 Cigarette demand curves in minimally nicotine dependent smokers (n=15; defined by a score of 1 or less on the Fagerstrom Test of Nicotine Dependence) and mildly to moderately nicotine dependent smokers (n=15; defined by a score of 2 or greater on the same measure). These curves were generated over the cigarette purchase task, with the *x* axis representing price in dollars ($) and the *y* axis representing self-reported consumption in cigarettes. Consistent with early work using linear demand analysis with the log of both axes, zero values were replaced with trivially small nonzero values (in this case, 0.001).

hypothetical task allows for the assessment of behaviors that may otherwise be hard to observe in laboratory settings for logistical reasons, but still may not reflect real-world choices as closely as later methods to assess demand.

Behavioral economics 3.0: naturalistic demand assessment

Naturalistic assessment of demand permits the purchase and use of tobacco products in daily life. This method provides solutions to the shortcomings of the hypothetical purchase tasks discussed above; most importantly, naturalistic assessment methods provide a medium to validate laboratory observations in an outpatient setting. To this end, naturalistic assessment gathers real-world evidence of purchasing and consumption that may not be easily or accurately mimicked in the laboratory.

Naturalistic demand assessment experiments typically allow participants to purchase products across a variety of prices and some, or all, of the products purchased are provided to the participant for consumption in the real world. Several important contributions have emerged from naturalistic demand assessment methodologies. In a preliminary study of the impact of price on nicotine gum, researchers found that brief physician counseling on the importance of quitting smoking in combination with a nicotine gum prescription (at $0, $6, $20 per box) impacted nicotine gum consumption and rates of cessation. Specifically, lower-priced gum increased the amount of gum purchased, amount of gum consumed, rate of long-term use, number of quit attempts, and number of people who maintained abstinence for one week (Hughes et al., 1991).

In a similar assessment of cigarette purchasing, participants were provided with an experimental income and instructed to purchase one week's worth of cigarettes under real and potentially real conditions. In the "real" condition, the participants completed four weekly sessions during which they purchased a week's worth of cigarettes that increased in price (i.e., $0.12, $0.25, $0.50, $1.00 per cigarette) each week. In the "potentially real" condition, during one session participants completed a cigarette purchase task where the cigarettes were available at four different prices. One of the four price conditions was selected at random to become real where the cigarettes purchased were provided by the experimenter for consumption over the next week. Regardless of condition, all participants completed a hypothetical cigarette purchase task. Finally, after each week, participants reported the number of cigarettes smoked over the course of the week. Importantly, Wilson et al. (2015) found that demand using the naturalistic and hypothetical purchase tasks was significantly different, but correlated. This indicates that purchasing in the naturalistic conditions was distinct from the hypothetical purchase task but still showed a similar pattern of decision-making.

Based on findings from Wilson et al. (2015), a subsequent study from the same laboratory investigated naturalistic demand across tobacco products (Stein et al., 2017). Products available for purchase now included either snus (a smokeless tobacco product) or nicotine gum, in addition to cigarettes. As with the previous study, one of the purchase task conditions was selected at random and the products purchased in that task were provided to the participant to consume over the course of one week (Stein et al., 2017). This study found differential effects of demand elasticity between snus, gum, and cigarettes such that snus had greater demand elasticity than cigarettes and gum. This may suggest that snus has lower abuse liability than cigarettes and gum.

Another study assessing naturalistic demand manipulated the effects of income on cigarette purchasing. Participants were experimentally provided an income and then completed four potentially real cigarette purchase tasks ($0.12, $0.25, $0.50, and $1.00 per cigarette) of which one was chosen to be actualized and the participant received the cigarettes purchased in addition to the remaining income that was left unspent. As shown in Figure 3.3, the amount of cigarettes purchased was systematically related to both cigarette price and income condition such that increases in cigarette price and decreases in income were both associated with a decrease in cigarettes purchased (Koffarnus, Wilson, and Bickel, 2015). The systematic relationship between income and cigarettes purchased is consistent with economic and human laboratory research on the influence of natural price changes on cigarette purchasing (DeGrandpre, Bickel, Rizvi, and Hughes, 1993; Hersch, 2000; Townsend et al., 1994), but is not consistent with previous research using the hypothetical purchase task which has found income to have no relationship (MacKillop et al., 2008) or an inverse relationship (Murphy et al., 2011) with cigarettes purchased. This discrepancy suggests that the naturalistic task is sensitive to a broader range of factors, such as income, that may influence purchasing.

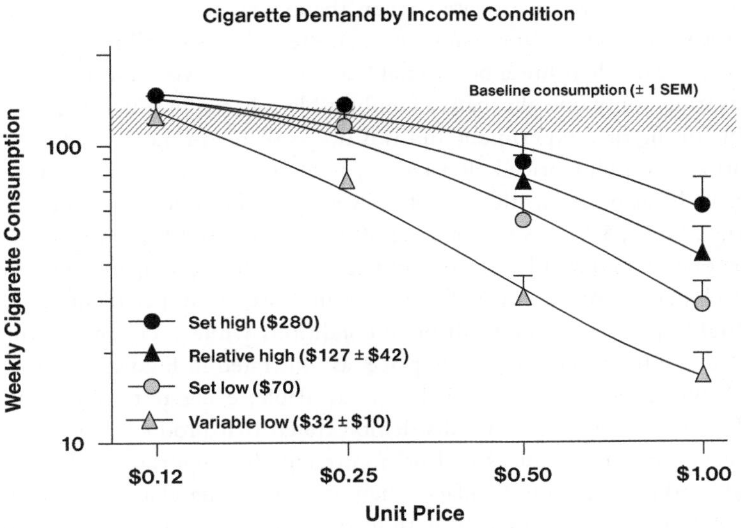

FIGURE 3.3 Both as the experimentally provided income decreases and as price for cigarettes increases weekly consumption of cigarettes decreases. (Reprinted from Koffarnus, Wilson, and Bickel, 2015).

Taken together, naturalistic demand assessment procedures provide methods for evaluating tobacco purchasing and consumption in a more real-world setting that can provide external validity for findings that were initially observed in the laboratory. Limitations of the studies described above are that the experimental manipulations exist in an open economy where there is limited constraint on other sources of tobacco purchases and consumption outside of the experiment; that few products have been assessed; and that the time-consuming nature of these studies makes it difficult to keep up with a quickly changing product landscape. The most recent method in the evolution of behavioral economics attempts to rectify some of these limitations.

Behavioral economics 4.0: experimental tobacco marketplace

The Experimental Tobacco Marketplace (ETM) was developed, in part, to account for the ever-changing product landscape and the time constraints imposed by the naturalistic assessment of demand discussed above. The ETM is an online analog to the complex tobacco marketplace, which was modeled from the experimental supermarket research that investigated the effects of tax and subsidy policies on food purchases (Epstein et al., 2012; Epstein et al., 2010). The ETM and experimental food supermarkets allow the experimenter to manipulate the type and number of available products, the information displayed, and the prices of all commodities rendering this method the most flexible and exhaustive of the behavioral economics paradigms for studying demand and substitution.

Using the ETM in the laboratory, the purchasing and consumption behavior of daily smokers was investigated when their current brand of cigarettes and other tobacco products (i.e., snus, e-cigarettes, dip, cigarillos, lozenges, and gum) were available (Quisenberry et al., 2015). In this two-experiment-within-subjects study, only one alternative combustible tobacco product, cigarillos, was available in the first experiment and was not available in the second, systematic replication. Constant-priced alternative products were available for purchase at a current, local price, while cigarettes were available under four price conditions (i.e., \$0.12, \$0.25, \$0.50, and \$1.00 per cigarette). Of the four price conditions, one was randomly selected and actualized for participants' use during a one-week period. As the price of cigarettes increased in Experiment 1, cigarillos were highly purchased across all price conditions while e-cigarette purchasing increased as a function of cigarette price, as illustrated in Figure 3.4. In Experiment 2, when cigarillos were no longer available, e-cigarettes and snus functioned as substitutes. These results demonstrated that product-type availability influenced consumer choices and demonstrated the importance of mimicking the real-world tobacco marketplace when studying demand and substitution.

The ETM can also be merged with features of the hypothetical purchase task and used as a data collection tool outside of the laboratory (i.e., for experiments conducted exclusively online) and in international samples. For example, in a sample of Norwegians, it took only eight minutes, on average, for participants to

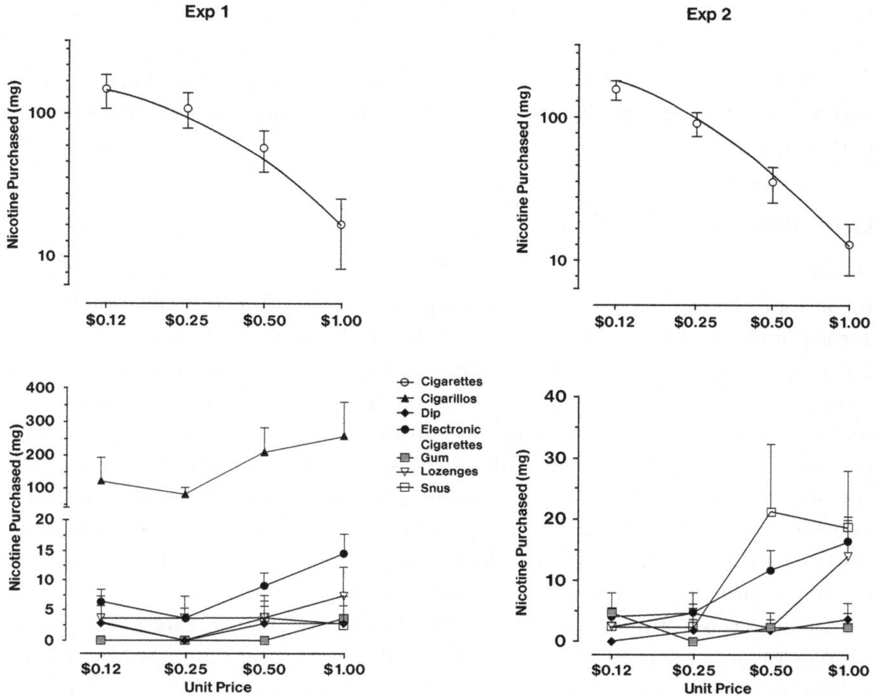

FIGURE 3.4 Data from the Experimental Tobacco Marketplace demonstrating substitution of different products dependent upon the available constant-priced products. In Experiment 1, cigarillos were purchased at all prices, while e-cigarettes were purchased as a function of increasing cigarette prices whereas in Experiment 2, when cigarillos were not available, e-cigarettes and snus substituted for cigarettes at high prices. (Reprinted from Quisenberry et al., 2015).

complete four completely hypothetical price conditions in the ETM (Heckman et al., 2016). When conventional or roll-your-own cigarettes were available at increasing prices, the constant-priced e-cigarette liquid tank systems functioned as strong substitutes, while de-nicotinized cigarettes functioned as weak substitutes. Follow-up conditions indicated that when e-cigarette liquid tank systems were not available, de-nicotinized cigarettes remained weak substitutes, unaffected by the availability of tank systems. A final condition determined that tank systems were an even stronger substitute when de-nicotinized cigarettes were not available. These results fortify that the complex interactions between product availability can determine the extent to which constant-priced products function as substitutes.

The ETM is both novel and the most comprehensive of the behavioral economics methods to study tobacco demand and substitution. Purchasing using this forum can be potentially real (Quisenberry et al., 2015) or purely hypothetical

(Heckman et al., 2016) and still produce informative results. Perhaps most importantly, the ETM can be used as a tool to prospectively study the rapidly transforming tobacco marketplace and provide information about new products and pricing strategies prior to these manipulations being implemented into the material world.

Limitations and ethics

Behavioral economics such as those explored here are not useful for all studies of tobacco products. For example, issues of quantifying the effects of market changes at the population level would not be suitable for a behavioral economics approach. Such questions would be better addressed by standard economics approaches. Other limitations of the specific behavioral economics methods discussed here include satiation during laboratory self-administration studies and practical limitations such as building suitable laboratory space. In hypothetical purchase task studies, the experimenter is reliant on self-report measures and in the naturalistic assessment of demand studies, manipulations exist in an open economy and these studies are time-consuming to conduct. Many of these limitations have fueled the development of assessment methodologies in subsequent epochs.

An important ethical consideration in conducting this research is exposing current tobacco users to novel tobacco products. Users of combustible tobacco products, unfortunately, are already exposing themselves to the most harmful form of tobacco products and one that is a leading cause of excess morbidity and mortality worldwide. Consistent with the Institute of Medicine's report (Institute of Medicine, 2012), exposing them to potentially less harmful, noncombustible products, in the context of a controlled experiment, informs decisions that could lead to better population health without increasing risk. Some special populations require additional consideration. For example, individuals who are seeking to quit or cut back on smoking, or adolescents who smoke, may not be ethically appropriate to include in studies where real tobacco products are provided. However, the hypothetical purchase task and hypothetical ETM methodologies are better suited for these special populations.

Conclusion

Collectively, these findings show the utility and power of behavioral economics for informing tobacco regulatory decisions. Over the last 25 years, this method has evolved to address the broad array of questions and conditions necessary to understand tobacco purchasing and consumption and has been useful in studying other forms of addiction. Along with the evolution of methods, the quantification of demand and substitution has also matured. These epochs of innovation, as discussed above, have ranged from very laborious, intensive laboratory studies to mimicking the actual tobacco marketplace. The results of these studies suggest that tobacco products can function as robust reinforcers that are sensitive to the broad range of market and environmental conditions. These studies may be used to contribute to tobacco regulatory science and inform public policy with the aim of improving public health.

References

Allison, J.W. (1983). *Behavioral Economics*. Westport, CT: Praeger Publishers.

Balster, R.L. (1991). "Drug Abuse Potential Evaluation in Animals," *British Journal of Addiction* 86(12): 1549–1558.

Bickel, W.K. and Madden, G. (1998). *The Behavioral Economics of Smoking*. Cambridge, MA: National Bureau of Economic Research. Available at http://www.nber.org/papers/w6444.pdf (accessed January 3, 2017).

Bickel, W.K., DeGrandpre, R.J., and Higgins, S.T. (1993). "Behavioral Economics: A Novel Experimental Approach to the Study of Drug Dependence," *Drug and Alcohol Dependence* 33(2): 173–192.

Bickel, W.K., DeGrandpre, R.J., and Higgins, S.T. (1995). "The Behavioral Economics of Concurrent Drug Reinforcers: A Review and Reanalysis of Drug Self-Administration Research," *Psychopharmacology* 118(3): 250–259.

Bickel, W.K., DeGrandpre, R.J., Hughes, J.R., and Higgins, S.T. (1991). "Behavioral Economics of Drug Self-Administration. II. A Unit-Price Analysis of Cigarette Smoking," *Journal of the Experimental Analysis of Behavior* 55(2): 145–154.

Bickel, W.K., Hughes, J.R., DeGrandpre, R.J., Higgins, S.T., and Rizzuto, P. (1992). "Behavioral Economics of Drug Self-Administration IV. The Effects of Response Requirement on the Consumption of and Interaction Between Concurrently Available Coffee and Cigarettes," *Psychopharmacology* 107, 211–216.

Carter, L.P. and Griffiths, R.R. (2009). "Principles of Laboratory Assessment of Drug Abuse Liability and Implications for Clinical Development," *Drug and Alcohol Dependence 105, Supplement* 1, S14–S25.

Carter, L.P., Stitzer, M.L., Henningfield, J.E., O'Connor, R.J., Cummings, K.M., and Hatsukami, D.K. (2009). "Abuse Liability Assessment of Tobacco Products Including Potential Reduced Exposure Products," *Cancer Epidemiology, Biomarkers & Prevention: A Publication of the American Association for Cancer Research, Cosponsored by the American Society of Preventive Oncology* 18(12): 3241–3262.

DeGrandpre, R.J., Bickel, W.K., Higgins, S.T., and Hughes, J.R. (1994). "A Behavioral Economic Analysis of Concurrently Available Money and Cigarettes," *Journal of the Experimental Analysis of Behavior* 61(2): 191–201.

DeGrandpre, R.J., Bickel, W.K., Hughes, J.R., and Higgins, S.T. (1992). "Behavioral Economics of Drug Self-Administration," *Psychopharmacology* 108(1–2): 1–10.

DeGrandpre, R.J., Bickel, W.K., Hughes, J.R., Layng, M.P., and Badger, G. (1993). "Unit Price as a Useful Metric in Analyzing Effects of Reinforcer Magnitude," *Journal of the Experimental Analysis of Behavior* 60(3): 641–666.

DeGrandpre, R.J., Bickel, W.K., Rizvi, S.A., and Hughes, J.R. (1993). "Effects of Income on Drug Choice in Humans," *Journal of the Experimental Analysis of Behavior* 59(3): 483–500.

Epstein, R. (1986). "Simulation Research in the Analysis of Behavior," in *Research Methods in Applied Behavior Analysis* (pp. 127–155). New York: Springer.

Epstein, L.H., Dearing, K.K., Roba, L.G., and Finkelstein, E. (2010). "The Influence of Taxes and Subsidies on Energy Purchased in an Experimental Purchasing Study," *Psychological Science* 21(3): 406–414.

Epstein, L.H., Jankowiak, N., Nederkoorn, C., Raynor, H.A., French, S.A., and Finkelstein, E. (2012). "Experimental Research on the Relation Between Food Price Changes and Food-Purchasing Patterns: A Targeted Review," *The American Journal of Clinical Nutrition* 95(4): 789–809.

Food and Drug Administration. (2010). "Guidance for Industry: Assessment of Abuse Potential of Drugs. Draft Guidance," *US Department of Health and Human Services, Center for Drug Evaluation and Research, Washington, DC.*

Giordano, L.A., Bickel, W.K., Shahan, T.A., and Badger, G.J. (2001). "Behavioral Economics of Human Drug Self-Administration: Progressive Ratio Versus Random Sequences of Response Requirements," *Behavioural Pharmacology* 12(5): 343–347.

Grace, R.C., Kivell, B.M., and Laugesen, M. (2015). "Estimating Cross-Price Elasticity of E-Cigarettes Using a Simulated Demand Procedure," *Nicotine & Tobacco Research: Official Journal of the Society for Research on Nicotine and Tobacco* 17(5): 592–598.

Guillem, K., Vouillac, C., Azar, M.R., Parsons, L.H., Koob, G.F., Cador, M., and Stinus, L. (2005). "Monoamine Oxidase Inhibition Dramatically Increases the Motivation to Self-Administer Nicotine in Rats," *The Journal of Neuroscience: The Official Journal of the Society for Neuroscience* 25(38): 8593–8600.

Heckman, B., Cummings, K.M., Heijndijk, S., Willemsen, M., Borland, R., O'Connor, R., and Bickel, W. (2016). "Substitutability of Reduced Nicotine Cigarettes and Electronic Cigarettes: The Experimental Tobacco Marketplace as a Tool to Inform Regulatory Standards for Novel Products," presented at the Society for Research on Nicotine & Tobacco, Chicago, IL.

Henningfield, J.E. and Keenan, R.M. (1993). "Nicotine Delivery Kinetics and Abuse Liability," *Journal of Consulting and Clinical Psychology* 61(5): 743–750.

Hersch, J. (2000). "Gender, Income Levels, and the Demand for Cigarettes," *Journal of Risk and Uncertainty* 21(2–3): 263–282.

Hughes, J.R., Wadland, W.C., Fenwick, J.W., Lewis, J., and Bickel, W.K. (1991). "Effect of Cost on the Self-Administration and Efficacy of Nicotine Gum: A Preliminary Study," *Preventive Medicine* 20(4): 486–496.

Hursh, S.R. (1980). "Economic Concepts for the Analysis of Behavior," *Journal of the Experimental Analysis of Behavior* 34(2): 219–238.

Hursh, S.R. (1984). "Behavioral Economics," *Journal of the Experimental Analysis of Behavior* 42(3): 435–452.

Hursh, S.R. and Roma, P.G. (2013). "Behavioral Economics and Empirical Public Policy," *Journal of the Experimental Analysis of Behavior* 99(1): 98–124.

Hursh, S.R. and Roma, P.G. (2015). "Behavioral Economics and the Analysis of Consumption and Choice," *Managerial and Decision Economics*. http://doi.org/10.1002/mde.2724.

Hursh, S.R. and Silberberg, A. (2008). "Economic Demand and Essential Value," *Psychological Review* 115(1): 186–198.

Hursh, S.R. and Winger, G. (1995). "Normalized Demand for Drugs and Other Reinforcers," *Journal of the Experimental Analysis of Behavior* 64(3): 373–384.

Hursh, S.R., Raslear, T.G., Bauman, R., and Black, H. (1989). "The Quantitative Analysis of Economic Behavior With Laboratory Animals," in *Understanding Economic Behaviour* (pp. 393–407). Netherlands: Springer.

Institute of Medicine. (2012). *Committee on Scientific Standards for Studies of Modified Risk Tobacco Products: Scientific Standards for Studies of Modified Risk Tobacco Products*. National Academies Press.

Jacobs, E.A. and Bickel, W.K. (1999). "Modeling Drug Consumption in the Clinic Using Simulation Procedures: Demand for Heroin and Cigarettes in Opioid-Dependent Outpatients," *Experimental and Clinical Psychopharmacology* 7(4): 412–426.

Johnson, M.W. and Bickel, W.K. (2003). "The Behavioral Economics of Cigarette Smoking: The Concurrent Presence of a Substitute and an Independent Reinforcer," *Behavioural Pharmacology* 14(2): 137–144.

Johnson, M.W., Bickel, W.K., and Kirshenbaum, A.P. (2004). "Substitutes for Tobacco Smoking: A Behavioral Economic Analysis of Nicotine Gum, Denicotinized Cigarettes, and Nicotine-Containing Cigarettes," *Drug and Alcohol Dependence* 74(3): 253–264.

Koffarnus, M.N., Franck, C.T., Stein, J.S., and Bickel, W.K. (2015). "A Modified Exponential Behavioral Economic Demand Model to Better Describe Consumption Data," *Experimental and Clinical Psychopharmacology* 23(6): 504–512.

Koffarnus, M.N., Wilson, A.G., and Bickel, W.K. (2015). "Effects of Experimental Income on Demand for Potentially Real Cigarettes," *Nicotine & Tobacco Research: Official Journal of the Society for Research on Nicotine and Tobacco* 17(3): 292–298.

MacKillop, J., Murphy, J.G., Ray, L.A., Eisenberg, D.T.A., Lisman, S.A., Lum, J.K., and Wilson, D.S. (2008). "Further Validation of a Cigarette Purchase Task for Assessing the Relative Reinforcing Efficacy of Nicotine in College Smokers," *Experimental and Clinical Psychopharmacology* 16(1): 57–65.

Madden, G.J. and Bickel, W.K. (1999). "Abstinence and Price Effects on Demand for Cigarettes: A Behavioral-Economic Analysis," *Addiction* 94(4): 577–588.

Murphy, J.G., MacKillop, J., Tidey, J.W., Brazil, L.A., and Colby, S.M. (2011). "Validity of a Demand Curve Measure of Nicotine Reinforcement with Adolescent Smokers," *Drug and Alcohol Dependence* 113(2–3): 207–214.

O'Brien, C.P. (1996). "Drug Addiction and Drug Abuse," *The Pharmacological Basis of Therapeutics* 10, 621–642.

O'Connor, R.J., June, K.M., Bansal-Travers, M., Rousu, M.C., Thrasher, J.F., Hyland, A., and Cummings, K.M. (2014). "Estimating Demand for Alternatives to Cigarettes with Online Purchase Tasks," *American Journal of Health Behavior* 38(1): 103–113.

Petry, N.M. and Bickel, W.K. (1998). "Polydrug Abuse in Heroin Addicts: A Behavioral Economic Analysis," *Addiction* 93(3): 321–335.

Quisenberry, A.J., Koffarnus, M.N., Hatz, L.E., Epstein, L.H., and Bickel, W.K. (2015). "The Experimental Tobacco Marketplace I: Substitutability as a Function of the Price of Conventional Cigarettes," *Nicotine & Tobacco Research: Official Journal of the Society for Research on Nicotine and Tobacco*. http://doi.org/10.1093/ntr/ntv230.

Shahan, T.A., Bickel, W.K., Badger, G.J., and Giordano, L.A. (2001). "Sensitivity of Nicotine-Containing and De-Nicotinized Cigarette Consumption to Alternative Non-Drug Reinforcement: A Behavioral Economic Analysis," *Behavioural Pharmacology* 12(4): 277–284.

Shahan, T.A., Bickel, W.K., Madden, G.J., and Badger, G.J. (1999). "Comparing the Reinforcing Efficacy of Nicotine Containing and De-Nicotinized Cigarettes: A Behavioral Economic Analysis," *Psychopharmacology* 147(2): 210–216.

Stafford, D., LeSage, M.G., and Glowa, J.R. (1998). "Progressive-ratio Schedules of Drug Delivery in the Analysis of Drug Self-Administration: A Review," *Psychopharmacology* 139(3): 169–184.

Stein, J.S., Koffarnus, M.N., Snider, S.E., Quisenberry, A.J., and Bickel, W.K. (2015). "Identification and Management of Nonsystematic Purchase Task Data: Toward Best Practice," *Experimental and Clinical Psychopharmacology* 23(5): 377.

Stein, J. S., Wilson, A. G., Koffarnus, M. N., Judd, M. C., & Bickel, W. K. (2017). "Naturalistic Assessment of Demand for Cigarettes, Snus, and Nicotine Gum." *Psychopharmacology* 234(2): 245–254.

Stitzer, M.L. and de Wit, H. (1998). "Abuse Liability," *Nicotine Safety and Toxicity*, 119.

Talhout, R., Opperhuizen, A., and van Amsterdam, J.G.C. (2007). "Role of Acetaldehyde in Tobacco Smoke Addiction," *European Neuropsychopharmacology: The Journal of the European College of Neuropsychopharmacology* 17(10): 627–636.

Townsend, J., Roderick, P., and Cooper, J. (1994). "Cigarette Smoking by Socio-economic Group, Sex, and Age: Effects of Price, Income, and Health Publicity," *BMJ* 309(6959): 923–927.

Wilson, A.G., Franck, C.T., Koffarnus, M.N., and Bickel, W.K. (2015). "Behavioral Economics of Cigarette Purchase Tasks: Within-Subject Comparison of Real, Potentially Real, and Hypothetical Cigarettes," *Nicotine & Tobacco Research: Official Journal of the Society for Research on Nicotine and Tobacco*. http://doi.org/10.1093/ntr/ntv154.

World Health Organization. (2008). "The Scientific Basis of Tobacco Product Regulation." Available at http://www.who.int/tobacco/global_interaction/tobreg/publications/9789241209519.pdf (accessed January 3, 2017).

4

UNDERSTANDING ALCOHOL AND OTHER DRUG USE VIA BEHAVIORAL ECONOMICS

Review and clinical applications

Michael Amlung, Joshua Gray, and James MacKillop

Introduction

Alcohol and drug abuse are among the leading causes of morbidity and mortality (Mokdad et al., 2004). Approximately half of Americans drink alcohol at least monthly, with 23 percent of adults engaging in binge drinking in the past month (SAMHSA, 2013). Data from the most recent wave of the National Epidemiologic Survey on Alcohol and Related Conditions (NESARC) study indicated that an estimated 32.6 million US adults met criteria for an alcohol use disorder within the last 12 months (Grant, Goldstein et al., 2015). Rates of other drug misuse are similarly high: an estimated 24.6 million Americans reported using drugs in the past month, with high prevalence rates of cannabis, prescription drug, and cocaine use (SAMHSA, 2013). In particular, abuse of prescription and non-prescription opioids has reached epidemic levels in the US and other nations (Manchikanti et al., 2012; Nelson and Perrone, 2012). The most recent NESARC data indicate nearly 4 percent of US adults met DSM-5 criteria for substance use disorder in the past year (Grant, Saha et al., 2015). The economic impact of alcohol and drug abuse is also substantial, with an estimated annual cost in the US exceeding $400 billion (Centers for Disease Control and Prevention, 2014; National Drug Intelligence Center, 2011). Finally, links between drug and alcohol misuse and criminal behavior are well documented (Boden et al., 2013; Popovici et al., 2012; Walters, 2014), with an estimated $108 billion spent annually on alcohol and drug-related criminal justice activities (Office of National Drug Control Policy, 2004).

Despite decades of research across a range of disciplines, understanding of the underlying etiology of these disorders and the most effective treatments remains incomplete. Translational research in behavioral economics has provided a novel approach to examining the core decision-making and reinforcement pathology that characterizes addictive disorders (Bickel et al., 2014; MacKillop, 2016; Vuchinich and Heather, 2003). Behavioral economics integrates concepts from psychology and microeconomics to understand the transactions that individuals make with the world. It provides a theoretical framework for investigating cost–benefit decision-making and how the balance between competing drug and non-drug reinforcers is disrupted in addiction. This chapter discusses applications of behavioral economics in the domain of alcohol and drug use with a focus on two constructs: delay discounting (DD), which is a measure of the level of preference for smaller-immediate rewards vs. larger-delayed reward, and substance demand, which is a measure of the relative value of alcohol and drugs. The chapter begins with a brief discussion of traditional economic approaches to addiction and a general overview of the concepts of behavioral economics demand and delay discounting. This is followed by a concise review of the relevant research in addiction. Next, promising clinical applications of behavioral economics in addiction are highlighted with an emphasis on future directions. The scope of this chapter is limited to alcohol and other drug use; the applications of behavioral economics to tobacco use are reviewed by Bickel et al. in Chapter 3 of this book.

Traditional economic approaches to addiction

A behavioral economics approach to alcohol and drug use should be clearly distinguished from more traditional applications of economics. There is an extensive body of distinct work investigating the relationships between economic factors and alcohol consumption. For example, there is extensive evidence that increases in alcohol taxation reduce population-level consumption (Wagenaar et al., 2009) and have significant positive consequences on alcohol-related disease/injury, violence, traffic crashes, sexually transmitted diseases, other drug use, and crime (Wagenaar et al., 2010). Similarly, there is robust evidence that minimum pricing reduces alcohol consumption (e.g., Stockwell et al., 2012) and reduces alcohol-related traffic violations and crimes against individuals (Stockwell et al., 2015). Other foci that fall within the application of economics to alcohol include studies on regulatory policy as it pertains to the supply of alcohol (i.e., varying market conditions from privatization to state monopolies, density of alcohol outlets, drink specials/"happy hours") (Treno et al., 2014; Xuan et al., 2015). Although these studies focus on the relations between microeconomic parameters of supply and demand in relation to drinking behavior, their focus is primarily on large-scale aggregate relationships rather than individual-level preferences. In contrast, behavioral economics studies focus on the choices, values, and preferences of individual people. That is not to say they are single-subject-design case studies (although that can be the case), but the critical distinction is that the unit of observation is data collected from individual people, not changes in unit sales, tax receipts, or other population-level indices. Finally, behavioral economics studies are commonly experimental in nature, whereas microeconomics studies are most often observational. Of note, compared to alcohol, traditional economic analysis is applied to a much smaller extent for illicit drugs because economic parameters, such as taxation, do not exist or are unregulated, but the same distinction applies.

Brief overview of behavioral economics research in addiction

Delay discounting

Delay discounting reflects the extent to which rewards are devalued based on their delay in time. It indexes preferences between smaller-immediate and larger-delayed rewards with greater immediate reward preference often construed as a form of impulsivity (Madden and Bickel, 2009). Delay discounting is commonly assessed via a series of dichotomous choices between immediate and delayed rewards (e.g., $40 today vs. $100 in a month). Although money is the most common reward, DD measures for drugs of abuse and health outcomes have proven useful in addictions research (e.g., Baker et al., 2003; Petry, 2001). An individual's rate of DD is typically quantified using a hyperbolic discounting function (i.e., k) that reflects the rate at which delayed rewards lose their value (Mazur, 1987). Larger k values reflect steeper discounting and greater impulsivity (see Figure 4.1A). Of note, other model-free methods are also used to quantify DD, including impulsive choice ratios (e.g., Mitchell, 1999) and area under the curve (e.g., Myerson et al., 2001). An overview of the different indices of DD is provided in Table 4.1.

Steep DD is a common clinical characteristic of addiction given that these individuals repeatedly seek smaller-immediate gains from alcohol and drug use over larger-delayed gains (e.g., financial stability, relationships, health). Across a large number of studies, individuals with addictive disorders have been shown to exhibit more precipitous DD relative to controls. This includes individuals with alcohol use disorders (e.g., Petry, 2001), cocaine dependence (e.g., Coffey et al., 2003), and opiate dependence (Madden et al., 1999). The notable exception to this pattern is cannabis, with prior research reporting no difference in DD between marijuana abusers and controls (Bidwell et al., 2013; Johnson et al., 2010; Mejía-Cruz et al., 2016). Significant correlations between steeper DD and higher addiction severity and greater quantity/frequency of use have been reported in several studies and with several substances (e.g., Kirby and Petry, 2004; MacKillop, Miranda et al., 2010; Petry, 2001).

A meta-analysis synthesized the findings of categorical studies comparing individuals exhibiting addictive behavior and controls (MacKillop et al., 2011). The meta-analytic sample comprised 57 comparisons, with >3,300 participants across six different categories of addictive behaviors (alcohol, tobacco, stimulant, cannabis, opiate, and gambling disorder). Across all studies, a highly significant, medium magnitude effect size (Cohen's $d = .62$) difference in DD was found between groups exhibiting addictive behavior and controls. Furthermore, significantly larger effect sizes were found for studies examining individuals meeting clinical criteria for substance use disorder ($d = .67$) compared to subclinical

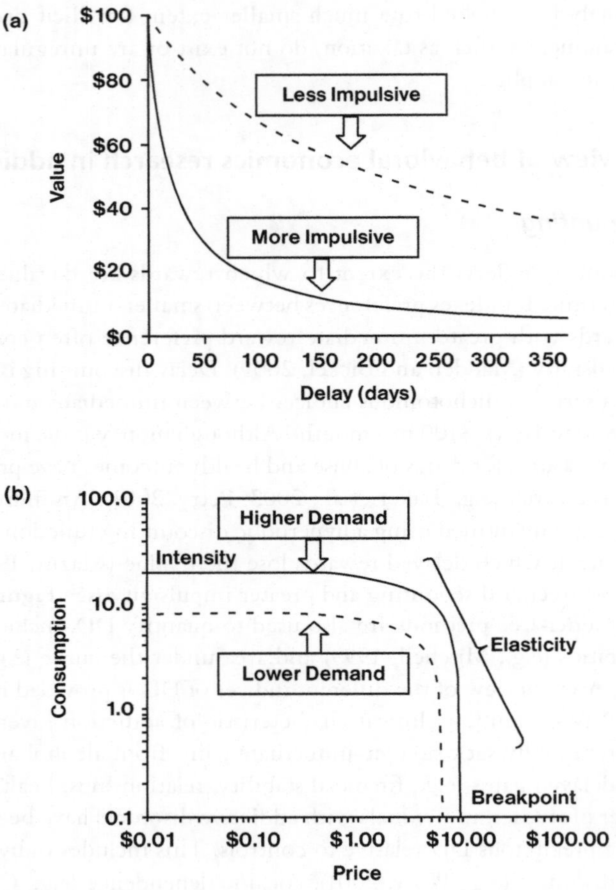

FIGURE 4.1 Prototypic delay discounting and demand curves. Panel A depicts hyperbolic delay discounting curves illustrating steep (more impulsive) and shallow (less impulsive) discounting in two hypothetical individuals. Panel B depicts demand curves illustrating the individual indices of demand and higher and lower levels in two hypothetical individuals.

samples of heavy or recreational users who do not meet clinical criteria (d = .46), and for studies using higher-resolution discounting measures compared to course assessments with only one or two items. Finally, the meta-analysis also suggested only modest influence of unpublished findings (i.e., publication bias) on the aggregated results. The results of this meta-analysis provide strong evidence that steep DD is a core feature of addictive behavior that is consistent across type of addiction and is meaningfully related to addiction severity.

The majority of prior research examining DD in samples of individuals with addiction has used cross-sectional designs which do not allow for determination of the causal relationship between steep DD and addiction (de Wit, 2009).

TABLE 4.1 Behavioral economics measures and associated indices of impulsivity and demand

Assessment/Index	*Description/Definition*
I. Delay Discounting Tasks	Behavioral economic measure of impulsivity; characterizes preferences for smaller immediate rewards over larger delayed rewards
k	Fitted parameter derived from hyperbolic discounting function that reflects overall rate of discounting of delayed rewards; larger *k* values reflect steeper discounting (i.e., greater impulsivity)
Impulsive choice ratio	Simple ratio of immediate reward choices to delayed reward choices; larger values reflect greater immediate reward preference (i.e., greater impulsivity).
Area under the curve (AUC)	Model-free index of delay discounting via geometric quantification of area under the discounting curve; AUC is inversely related to *k,* such that smaller values reflect steeper discounting (i.e., greater impulsivity)
II. Purchase Tasks (e.g., Alcohol Purchase Task; Marijuana Purchase Task; Cocaine Purchase Task)	Behavioral economic measure of substance demand; characterizes cost–benefit decision-making via estimated consumption at varying levels of price
Intensity	Consumption at zero price
Breakpoint	First price that suppresses consumption to zero
O_{max}	Maximum expenditure across prices
P_{max}	Price at which demand curve first becomes elastic; also price associated with O_{max}
Elasticity	Proportional slope of the demand curve

Retrospective studies have found that steeper DD in adolescence predicts earlier onset of alcohol use disorder symptoms (Dom et al., 2006; Kollins, 2003). Other studies have found that DD prospectively predicts severity of alcohol use over a six-month period (Fernie et al., 2013). However, some studies have found that the DD rate may become less impulsive following abstinence. Both ex- and never smokers have been shown to exhibit less steep discounting for monetary rewards relative to current smokers (Bickel et al., 1999). Another study found that discounting rates for delayed health outcomes among ex-smokers were between current and never smokers, although not significantly different from either group (Odum et al., 2002). Finally, a recent longitudinal study found that smokers who maintained abstinence across a 12-month period following a smoking cessation intervention exhibited significant decreases in monetary DD, while no change occurred in individuals who resumed smoking (Secades-Villa et al., 2014). However, DD rates were not affected by a brief alcohol and drug use intervention in college student drinkers (Dennhardt et al., 2015). These studies offer mixed support for reversal of steep DD following abstinence, with the most promising data to date pertaining primarily to nicotine dependence. Further understanding the causal relations between DD and addiction is an important priority and may help clarify the role of DD in preventing the onset of addictive disorders as well as developing more effective treatments (see below).

Demand

A common clinical feature of alcohol and drug addiction is the persistent overvaluation of alcohol and drug rewards relative to other reinforcers (Bickel et al., 2014). This overvaluation is captured via the behavioral economics concept of demand, which reflects how much of a commodity is consumed at a given price. Historically, demand was typically measured via operant self-administration paradigms (Higgins et al., 1994; Mello and Mendelson, 1965), but recent research in human samples has led to the development of purchase tasks in which participants estimate consumption at varying levels of price. Purchase tasks have been developed to assess demand for alcohol (Murphy and MacKillop, 2006), cigarettes (Jacobs and Bickel, 1999; MacKillop et al., 2008), cocaine (Bruner and Johnson, 2014), and marijuana (Collins et al., 2014). These measures ask participants to report how many units of the commodity (e.g., number of alcoholic drinks, marijuana joints, or hits of cocaine) they would purchase and consume at a range of prices (e.g., "How many drinks would you consume if they cost $4 each?"). Plotting consumption by price yields a demand curve (Figure 4.1B) that can be used to generate five conceptually related indices of demand (Table 4.1). Of note, the axes in Figure 4.1B are reversed from the typical demand curve format initially presented by Marshall (Friedman, 1949; Marshall, 1898); however, the convention in psychology is to plot price along the x-axis and consumption on the y-axis. Furthermore, in psychological applications of behavioral economics, the elasticity index (Table 4.1) is viewed as a measure of proportional price sensitivity or overall slope of the demand curve (e.g., Hursh and Silberberg, 2008), which differs somewhat from traditional economic definitions of elasticity. Detailed descriptions of equations used to model elasticity from behavioral economics purchase task data are provided in Hursh and Silberberg (2008) and Koffarnus et al. (2015).

Prior research examining demand for addictive substances has focused on two primary methodologies. The first is trait-based demand, reflecting estimated typical level of demand, and the second is state-based demand, reflecting estimated consumption in specific conditions or situations. In the case of trait-based demand, a growing literature has revealed robust associations between elevated demand for alcohol and drugs and quantity/frequency of use and clinical severity. For instance, higher alcohol demand in alcohol purchase tasks has been consistently associated with greater quantity/frequency of alcohol consumption, heavy drinking, and alcohol use disorder severity (Bertholet et al., 2015; MacKillop, Miranda et al., 2010; Murphy and MacKillop, 2006). In the case of drug demand, increased cocaine demand was shown to be correlated with greater cocaine use in cocaine-dependent individuals (Bruner and Johnson, 2014), and elevated marijuana demand has been shown to be related to greater marijuana use and symptoms of marijuana dependence (Aston et al., 2015; Collins et al., 2014).

The second domain of research in this area has focused on state-based influences on demand. To date, the main focus of this work has been alcohol demand, with a relatively smaller number of studies on tobacco demand (Acker and MacKillop, 2013; MacKillop et al., 2012). In the context of alcohol demand, one finding that has been replicated across a number of studies is that the relative value of alcohol is increased by alcohol-related environmental cues. Specifically, MacKillop, O'Hagen et al. (2010) observed significant increases in several indices of alcohol demand in heavy drinkers following a laboratory alcohol cue exposure. These findings were replicated in a subsequent study (Amlung et al., 2012) that also demonstrated close correspondence between estimated demand on an alcohol purchase task and actual alcohol consumption during a laboratory self-administration protocol. In another set of studies, negative affect and stress inductions significantly increased alcohol demand in heavy drinkers (Amlung and MacKillop, 2014; Owens, Ray, and MacKillop, 2015). Finally, Amlung et al. (2015) recently demonstrated that demand for alcohol is dynamically increased by acute alcohol consumption, which may contribute to loss of control over drinking that can occur following intoxication. These findings suggest that state-based indices of demand may complement existing measures of alcohol motivation, such as subjective craving. Importantly, no studies have examined state-based influences on demand for illicit drugs, which is an important target for future research.

Clinical applications of behavioral economics

As indicated by the preceding literature review, behavioral economics has emerged as a useful framework for investigating the underlying reinforcement pathology that characterizes alcohol and drug addiction. The logical extension of this research involves exploring the clinical applications of behavioral economics in the context of addiction. This section reviews evidence that measures of DD and demand can be used prospectively to predict the likelihood of treatment success, and provides an overview of novel behavioral economics-based interventions for addiction that focus on changing individuals' levels of DD and demand.

Predicting treatment response

High rates of relapse following treatment for alcohol and other drug use disorders have been well documented. For instance, two large systematic reviews of relapse rates following alcohol treatment found that between 66–79 percent of individuals with alcohol use disorders experience at least one relapse following treatment (Miller et al., 2001; Moyer and Finney, 2002). The high percentage of individuals who return to problematic use of alcohol and drugs following treatment underscores the importance of identifying risk factors for relapse and improving current treatment approaches. An increasing number of studies have examined whether DD and demand can be used to predict treatment response.

Several studies have shown that impulsive DD is associated with poor addiction treatment outcomes, though much of this work has focused on nicotine dependence. One study found that steeper DD was associated with a shorter number of days to first lapse following smoking cessation treatment (MacKillop and Kahler, 2009). Individuals who failed to maintain abstinence at two- and eight-week follow-up visits were also significantly more impulsive at baseline compared to those who were successful. Similar results have been reported in other studies of smokers (Dallery and Raiff, 2007; Krishnan-Sarin et al., 2007; Sheffer et al., 2012; Yoon et al., 2007). To date, two studies have examined whether DD predicts treatment outcomes in drug abusers. Washio et al. (2011) found that steeper DD is associated with worse treatment response following outpatient treatment for cocaine dependence. In another study, Stanger et al. (2012) found that impulsive DD at baseline prospectively predicted treatment outcomes in adolescents with marijuana use disorder.

The literature examining whether level of demand is associated with treatment outcomes is limited to a handful of studies focused on alcohol and tobacco use. MacKillop and Murphy (2007) found that indices of alcohol demand predicted post-treatment drinking behavior in college student drinkers who completed a brief alcohol intervention. Murphy et al. (2015) examined whether baseline performance on an alcohol purchase task predicted drinking levels and alcohol problems at 1 and 6 months following brief alcohol interventions in college drinkers. Elevated alcohol demand predicted increased drinks per week and greater alcohol problems at 1-month follow up. Furthermore, greater reductions in demand immediately following the intervention were associated with better outcomes at 1 month, but not at 6 months. In the case of nicotine dependence, elevated demand for cigarettes was associated with fewer abstinent carbon monoxide readings among smokers receiving brief advice and nicotine patch smoking-cessation treatment in a residential substance use treatment setting (MacKillop et al., 2015). However, the indices of demand did not predict abstinence at 1- and 3-month follow-ups.

Taken together, these studies indicate that behavioral economics measures of DD and substance demand (at least for alcohol and tobacco) may have clinical utility as predictors of risk for relapse following treatment. For DD, this has several potentially important clinical applications. First, impulsive DD may serve as a target for novel treatments for addiction, as described below. Second, assessment of DD may highlight individuals who are at particularly high risk for relapse, and may inform enhanced and individualized treatments for these individuals. Third, impulsive DD could indicate a positive prognosis and suggest a lower risk for relapse. Ultimately, these possibilities are largely speculative and are worthwhile pursuits for future research. The clinical applications for demand are less clear, given the relatively limited research conducted to date. While demand has been shown to predict early response to substance use treatment, its long-term predictive power has been mixed. Demand measures may also have clinical utility in identifying abuse liability of novel tobacco products (see Chapter 3 by Bickel et al. in this book).

Well-established reinforcement-based interventions for addiction

Clinical applications of behavioral economics often focus on increasing the value of abstinence and reducing the value of substance use by altering an individual's reinforcement contingencies. One well-established reinforcement-based treatment is the community reinforcement approach (CRA; Hunt and Azrin, 1973). This involves working with the individual to rearrange their vocational, social, recreational, and familial contingencies in order to restore healthy forms of reinforcement that are distinct from substance use (Smith et al., 2001). The CRA is designed to break the vicious cycle commonly seen in substance use disorders whereby heavy substance use leads to negative outcomes (e.g., loss of occupation or personal relationships), which in turn leads to reduced reinforcement from activities other than substance use. A systematic review of research on CRA found strong support for its efficacy for cocaine and alcohol use disorders, but somewhat limited efficacy for opioid dependence (Roozen et al., 2004). In addition, two recent randomized trials have provided initial support for implementing CRA as an internet-based treatment (Campbell et al., 2014; Christensen et al., 2014). A final adaptation of the CRA that specifically engages family members (the community reinforcement approach family treatment; CRAFT) has been shown to be effective in alcohol (e.g., Dutcher et al., 2009) and other substance use disorders (e.g., Brigham et al., 2014).

The second well-established form of reinforcement-based treatment is contingency management (CM). The CM approach differs from CRA in that it focuses more directly on reinforcing pro-treatment outcomes using incentives (Stitzer and Petry, 2006). Generally speaking, CM involves providing tangible rewards to reinforce positive behaviors such as abstinence (National Institute on Drug Abuse, 2012). In the context of alcohol use disorders, this might include providing a cash reward or voucher that can be redeemed for food items, movie passes, etc. each time an individual has a negative biochemical screen for recent alcohol use. Therefore, the CM approach aims to shift reinforcement away from alcohol or drug use toward behaviors that are consistent with a drug-free lifestyle. This approach has been shown to be effective in treating alcohol and other substance use disorders, and it has been implemented successfully by trained community providers (Petry et al., 2012; Petry et al., 2000; Prendergast et al., 2006).

A number of randomized clinical trials have demonstrated the efficacy of a number of variants of CM, including internet delivery and with the use of non-monetary and non-voucher-based reinforcement (e.g., Prendergast et al., 2006; Schumacher et al., 2007). By comparison, the number of randomized controlled trials of the CRA approach remains limited. Additional research is needed to verify the efficacy of CRA across a range of mediums (e.g., in person or internet-delivered; family involvement), types of addictive disorders, and populations. Furthermore, the underlying mechanisms of action of both of these interventions have yet to be fully characterized (e.g., does a change in DD or substance demand mediate the relationship between CRA and CM and positive outcomes?).

Novel interventions to reduce demand or impulsive discounting in addiction

Recent research has begun to explore novel behavioral-economics-based interventions that seek to directly alter individuals' level of DD or increase substance-free reinforcement. A variety of experimental manipulations have been shown to increase the salience of delayed rewards, thereby decreasing DD (for a review, see Koffarnus et al., 2013). For example, episodic future thinking has been shown to significantly reduce impulsive DD compared to a control imagery condition (Daniel et al., 2013; Daniel et al., 2015; Peters and Buchel, 2010). Furthermore, a study of individuals with alcohol use disorder suggest that episodic future thinking reduced DD rates and reduced hypothetical alcohol demand (Snider et al., 2016). Finally, two studies have shown that having participants interact with age-progressed computer-generated images of themselves reduced DD and risky long-term health-related decisions (Hershfield et al., 2011; Kaplan et al., 2015). While shifting mental orientation farther into the future successfully altered DD in these studies, whether the effect on DD is long-lasting remains unclear.

A second strategy for altering DD relies on working memory training as a way to strengthen the brain's executive system which is thought to be hypoactive in addicted individuals (Bickel et al., 2007). Specifically, Bickel et al. (2011) randomly assigned a sample of treatment-seeking stimulant addicts to receive either a working memory training regimen or control training of similar duration. On average, the active training group showed a 50% reduction in DD rates following the working memory training. If replicated, these results could indicate a novel strategy for reducing impulsive DD; however, as with the future forecasting techniques, additional studies are needed to determine whether these effects have any influence on substance-related variables.

An emerging line of research has explored the effectiveness of pairing traditional brief motivational interventions (BMIs) with behavioral-economics-based supplements (for a review, see Murphy et al., 2007). To date, the behavioral economics supplement that has shown the most promise is the Substance Free Activity Session (SFAS), developed by Murphy and colleagues (Murphy, Dennhardt et al., 2012; Murphy, Skidmore et al., 2012). The SFAS supplement consists of a 50-minute individual session that is rooted in a motivational interviewing framework. The supplement attempts to increase engagement in constructive, alcohol-free activities, draw attention to the potentially negative relationship between alcohol use and long-term goal accomplishment, and increase the salience of delayed rewards that are personally relevant to the college student population (e.g., academic and career success). The efficacy of the SFAS supplement has been evaluated in two studies, including an initial pilot study (Murphy, Skidmore et al., 2012) and a subsequent randomized controlled trial (Murphy, Dennhardt et al., 2012). Both studies reported significant improvement in alcohol-related outcomes (i.e., alcohol problems, heavy drinking) in the combined BMI and supplement group compared to the BMI alone group. A more recent randomized controlled trial provided initial evidence of the efficacy of an abbreviated SFAS in college student marijuana users (Yurasek et al., 2015). Further research is needed to evaluate the ideal duration and timing of these supplements, and to evaluate their utility for other substances.

Challenges and ethical issues

Despite considerable progress made from behavioral economics applications to addictive disorders, this approach has also faced challenges. In many cases, the measures used to assess DD or substance demand are based on self-report and hypothetical rewards. There is a risk that individuals' responses on these measures may not reflect their actual behavior if real rewards were at stake. However, a number of studies to date have demonstrated close correspondence between DD measures for hypothetical and incentivized rewards (e.g., Johnson and Bickel, 2002; Madden et al., 2003). In addition, the neurobiological substrates of DD do not appear to differ significantly between hypothetical and actual rewards (Bickel et al., 2009). In the case of demand, two studies have found that self-reported consumption on hypothetical and incentivized alcohol purchase tasks are highly correlated (Amlung et al., 2012; Amlung and MacKillop, 2015). Self-reported consumption on alcohol and cigarette purchase tasks is also highly correlated with actual drinking and smoking during laboratory self-administration protocols (Amlung et al., 2012; Amlung and MacKillop, 2015; MacKillop et al., 2012).

A second challenge that may limit the broad implementation of behavioral economics measures into large, population-based research studies is that the duration of some of the assessments may be prohibitive. Conveniently, behavioral economics researchers have begun to develop abbreviated DD and demand measures that allow for efficient and valid assessment of preferences. These include five-to-eight item discounting measures (Gray et al., 2014; Koffarnus and Bickel, 2014) and brief demand assessments (Amlung et al., 2015; Owens, Murphy, and MacKillop, 2015). Additional research is needed to further validate and extend the application of these abbreviated measures.

A few ethical issues are also important to consider when using behavioral economics assessments in samples of individuals with addiction. For instance, when studying individuals who are in treatment or recovery, there are clear ethical concerns with providing actual outcomes on substance demand measures (e.g., providing alcoholic drinks to an individual in treatment for alcohol problems). Doing so would also be against the ethical guidelines on ethanol administration provided by the NIAAA (National Institute on Alcohol Abuse and Alcoholism, 2005). However, as indicated above, hypothetical demand measures have been shown to provide valid assessments of actual drinking and smoking behavior and can be used in these contexts. Another potential ethical issue involves providing monetary incentives from DD and demand measures to individuals with addictive disorders due to concerns that the money may be used to acquire alcohol or drugs. However, a number of studies have found that providing cash incentives in research studies does not significantly instigate subsequent drug use (e.g., Festinger and Dugosh, 2012).

Conclusion

The goal of this chapter was to provide a concise review of behavioral economics research exploring DD and substance demand in the context of alcohol and other drug misuse. The empirical literature reviewed here suggests that individuals who abuse alcohol and other drugs consistently exhibit steeper discounting of delayed rewards and elevated substance demand relative to controls. Furthermore, these constructs appear to have clinical utility as predictors of treatment outcomes. Finally, there are a number of emerging clinical applications of behavioral economics that have yielded promising results in a handful of studies. However, in many cases, existing studies have focused exclusively on one type of addictive disorder (e.g., treatments for alcohol use disorder), or have predominantly been employed in non-addiction samples. Thus, a priority for the future is continuing to expand the use of these interventions across a range of substances and populations.

Although the literature on the behavioral economics of addiction has grown substantially over the past few decades, several important research priorities remain. First, as noted previously, the majority of the research conducted to date has used cross-sectional designs that preclude drawing conclusions about the developmental role of DD and demand in addiction. There is a clear need for longitudinal research to address whether demand and DD are causes of addiction, consequences of substance use, or a combination of both (de Wit, 2009). Second, the number of randomized controlled trials remains relatively small. Finally, two emerging research areas that have particular promise in addiction involve studying the underlying neurobiology of DD and demand via neuroeconomics (e.g., Amlung et al., 2014; Boettiger et al., 2007; MacKillop et al., 2014; Monterosso et al., 2007) and examining the genetic basis of these constructs (e.g., MacKillop and Munafo, 2013). Ultimately, though additional research is still needed in a number of areas, the literature reviewed in this chapter provides strong support for the utility of behavioral economics as a translational framework for research on addiction.

Acknowledgments

Dr. MacKillop's contributions partially supported by the Peter Boris Chair in Addictions Research and NIH grant P30 DA027827. Dr. Amlung's contributions partially supported by the Peter Boris Centre for Addictions Research at McMaster University and St. Joseph's Healthcare Hamilton.

References

Acker, J. and MacKillop, J. (2013). "Behavioral Economic Analysis of Cue-Elicited Craving for Tobacco: A Virtual Reality Study," *Nicotine & Tobacco Research* 15: 1409–1416.

Amlung, M. and MacKillop, J. (2014). "Understanding the Effects of Stress and Alcohol Cues on Motivation for Alcohol via Behavioral Economics," *Alcoholism, Clinical and Experimental Research* 38: 1780–1789.

Amlung, M. and MacKillop, J. (2015). "Further Evidence of Close Correspondence for Alcohol Demand Decision Making for Hypothetical and Incentivized Rewards." *Behavioural Processes* 113: 187–191.

Amlung, M., Acker, J., Stojek, M.K., Murphy, J.G., and MacKillop, J. (2012). "Is Talk 'Cheap'? An Initial Investigation of the Equivalence of Alcohol Purchase Task Performance for Hypothetical and Actual Rewards," *Alcoholism, Clinical and Experimental Research* 36: 716–724.

Amlung, M., McCarty, K.N., Morris, D.H., Tsai, C.-L., and McCarthy, D.M. (2015). "Increased Behavioral Economic Demand and Craving for Alcohol Following a Laboratory Alcohol Challenge," *Addiction* 110(9): 1421–1428.

Amlung, M., Sweet, L.H., Acker, J., Brown, C.L., and MacKillop, J. (2014). "Dissociable Brain Signatures of Choice Conflict and Immediate Reward Preferences in Alcohol Use Disorders," *Addiction Biology* 19: 743–753.

Aston, E.R., Metrik, J., and MacKillop, J. (2015). "Further Validation of a Marijuana Purchase Task," *Drug and Alcohol Dependence* 152: 32–38.

Baker, F., Johnson, M.W., and Bickel, W.K. (2003). "Delay Discounting in Current and Never-Before Cigarette Smokers: Similarities and Differences Across Commodity, Sign, and Magnitude," *Journal of Abnormal Psychology* 112: 382–392.

Bertholet, N., Murphy, J.G., Daeppen, J.-B., Gmel, G., and Gaume, J. (2015). "The Alcohol Purchase Task in Young Men from the General Population," *Drug and Alcohol Dependence* 146: 39–44.

Bickel, W.K., Johnson, M.W., Koffarnus, M.N., MacKillop, J., and Murphy, J.G. (2014). "The Behavioral Economics of Substance Use Disorders: Reinforcement Pathologies and Their Repair," *Annual Review of Clinical Psychology* 10: 641–677.

Bickel, W.K., Miller, M.L., Yi, R., Kowal, B.P., Lindquist, D.M., and Pitcock, J.A. (2007). "Behavioral and Neuroeconomics of Drug Addiction: Competing Neural Systems and Temporal Discounting Processes," *Drug and Alcohol Dependence* 90 (Suppl 1): S85–91.

Bickel, W.K., Odum, A.L., and Madden, G.J. (1999). "Impulsivity and Cigarette Smoking: Delay Discounting in Current, Never, and Ex-Smokers," *Psychopharmacology (Berl)* 146: 447–454.

Bickel, W.K., Pitcock, J.A., Yi, R., and Angtuaco, E.J.C. (2009). "Congruence of BOLD Response Across Intertemporal Choice Conditions: Fictive and Real Money Gains and Losses." *Journal of Neuroscience* 29: 8839–8846.

Bickel, W.K., Yi, R., Landes, R.D., Hill, P.F., and Baxter, C. (2011). "Remember the Future: Working Memory Training Decreases Delay Discounting Among Stimulant Addicts," *Biological Psychiatry* 69: 260–265.

Bidwell, L.C., Metrik, J., McGeary, J., Palmer, R.H.C., Francazio, S., and Knopik, V.S. (2013). "Impulsivity, Variation in the Cannabinoid Receptor (CNR1) and Fatty Acid Amide Hydrolase (FAAH) Genes, and Marijuana-Related Problems," *Journal of Studies on Alcohol and Drugs* 74: 867–878.

Boden, J.M., Fergusson, D.M., and Horwood, L.J. (2013). "Alcohol Misuse and Criminal Offending: Findings from a 30-Year Longitudinal Study," *Drug and Alcohol Dependence* 128: 30–36.

Boettiger, C.A., Mitchell, J.M., Tavares, V.C., Robertson, M., Joslyn, G., D'Esposito, M., et al. (2007). "Immediate Reward Bias in Humans: Fronto-Parietal Networks and a Role for the Catechol-O-Methyltransferase 158(Val/Val) Genotype," *Journal of Neuroscience* 27: 14383–14391.

Brigham, G.S., Slesnick, N., Winhusen, T.M., Lewis, D.F., Guo, X., and Somoza, E. (2014). "A Randomized Pilot Clinical Trial to Evaluate the Efficacy of Community Reinforcement and Family Training for Treatment Retention (CRAFT-T) for Improving Outcomes for Patients Completing Opioid Detoxification," *Drug and Alcohol Dependence* 138: 240–243.

Bruner, N.R. and Johnson, M.W. (2014). "Demand Curves for Hypothetical Cocaine in Cocaine-Dependent Individuals," *Psychopharmacology* 231: 889–897.

Campbell, A.N.C., Nunes, E.V., Matthews, A.G., Stitzer, M., Miele, G.M., Polsky, D., et al. (2014). "Internet-Delivered Treatment for Substance Abuse: A Multisite Randomized Controlled Trial," *American Journal of Psychiatry* 171: 683–690.

Centers for Disease Control and Prevention (2014). *Excessive Drinking Costs U.S. $223.5 Billion*. Available at http://www.cdc.gov/features/alcoholconsumption/ (accessed November 15, 2016).

Christensen, D.R., Landes, R.D., Jackson, L., Marsch, L.A., Mancino, M.J., Chopra, M.P., et al. (2014). "Adding an Internet-Delivered Treatment to an Efficacious Treatment Package for Opioid Dependence," *Journal of Consulting and Clinical Psychology* 82: 964–972.

Coffey, S.F., Gudleski, G.D., Saladin, M.E., and Brady, K.T. (2003). "Impulsivity and Rapid Discounting of Delayed Hypothetical Rewards in Cocaine-Dependent Individuals," *Experimental and Clinical Psychopharmacology* 11: 18–25.

Collins, R.L., Vincent, P.C., Yu, J., Liu, L., and Epstein, L.H. (2014). "A Behavioral Economic Approach to Assessing Demand for Marijuana," *Experimental and Clinical Psychopharmacology* 22: 211–221.

Dallery, J. and Raiff, B.R. (2007). "Delay Discounting Predicts Cigarette Smoking in a Laboratory Model of Abstinence Reinforcement," *Psychopharmacology* 190: 485–496.

Daniel, T.O., Said, M., Stanton, C.M., and Epstein, L.H. (2015). "Episodic Future Thinking Reduces Delay Discounting and Energy Intake in Children," *Eating Behaviors* 18: 20–24.

Daniel, T.O., Stanton, C.M., and Epstein, L.H. (2013). "The Future Is Now: Reducing Impulsivity and Energy Intake Using Episodic Future Thinking," *Psychological Science* 24: 2339–2342.

Dennhardt, A.A., Yurasek, A.M., and Murphy, J.G. (2015). "Change in Delay Discounting and Substance Reward Value Following a Brief Alcohol and Drug Use Intervention," *Journal of the Experimental Analysis of Behavior* 103: 125–140.

De Wit, H. (2009). "Impulsivity as a Determinant and Consequence of Drug Use: A Review of Underlying Processes," *Addiction Biology* 14: 22–31.

Dom, G., D'Haene, P., Hulstijn, W., and Sabbe, B. (2006). "Impulsivity in Abstinent Early- and Late-Onset Alcoholics: Differences in Self-Report Measures and a Discounting Task," *Addiction* 101: 50–59.

Dutcher, L.W., Anderson, R., Moore, M., Luna-Anderson, C., Meyers, R., Delaney, H.D., et al. (2009). "Community Reinforcement and Family Training (CRAFT): An Effectiveness Study," *Journal of Behavior Analysis in Health, Sports, Fitness and Medicine* 2: 80–93.

Fernie, G., Peeters, M., Gullo, M.J., Christiansen, P., Cole, J.C., Sumnall, H., et al. (2013). "Multiple Behavioural Impulsivity Tasks Predict Prospective Alcohol Involvement in Adolescents," *Addiction (Abingdon, England)* 108: 1916–1923.

Festinger, D.S. and Dugosh, K.L. (2012). "Paying Substance Abusers in Research Studies: Where Does the Money Go?" *The American Journal of Drug and Alcohol Abuse, 38,* 43–48.

Friedman, M. (1949). "The Marshallian Demand Curve," *Journal of Political Economy* 57: 463–495.

Grant, B.F., Goldstein, R.B., Saha, T.D., Chou, S.P., Jung, J., Zhang, H., et al. (2015). "Epidemiology of *DSM-5* Alcohol Use Disorder," *JAMA Psychiatry* 72: 757–766.

Grant, B.F., Saha, T.D., Ruan, W.J., Goldstein, R.B., Chou, S.P., Jung, J., et al. (2015). "Epidemiology of DSM-5 Drug Use Disorder," *JAMA Psychiatry* 73: 39–47.

Gray, J.C., Amlung, M.T., Acker, J.D., Sweet, L.H., and MacKillop, J. (2014). "Item-based Analysis of Delayed Reward Discounting Decision Making," *Behavioural Processes* 103: 256–260.

Hershfield, H.E., Goldstein, D.G., Sharpe, W.F., Fox, J., Yeykelis, L., Carstensen, L.L., et al. (2011). "Increasing Saving Behavior Through Age-Progressed Renderings of the Future Self," *Journal of Marketing Research* 48: S23–S37.

Higgins, S.T., Bickel, W.K., and Hughes, J.R. (1994). "Influence of an Alternative Reinforcer on Human Cocaine Self-Administration," *Life Sciences* 55: 179–187.

Hunt, G.M. and Azrin, N.H. (1973). "A Community-Reinforcement Approach to Alcoholism," *Behaviour Research and Therapy* 11: 91–104.

Hursh, S.R. and Silberberg, A. (2008). "Economic Demand and Essential Value," *Psychological Review* 115: 186–198.

Jacobs, E.A. and Bickel, W.K. (1999). "Modeling Drug Consumption in the Clinic Using Simulation Procedures: Demand for Heroin and Cigarettes in Opioid-Dependent Outpatients," *Experimental and Clinical Psychopharmacology* 7: 412–426.

Johnson, M.W. and Bickel, W.K. (2002). "Within-Subject Comparison of Real and Hypothetical Money Rewards in Delay Discounting." *Journal for the Experimental Analysis of Behavior* 77: 129–146.

Johnson, M.W., Bickel, W.K., Baker, F., Moore, B.A., Badger, G.J., and Budney, A.J. (2010). "Delay Discounting in Current and Former Marijuana-Dependent Individuals," *Experimental and Clinical Psychopharmacology* 18: 99–107.

Kaplan, B.A., Reed, D.D., and Jarmolowicz, D.P. (2015). "Effects of Episodic Future Thinking on Discounting: Personalized Age-Progressed Pictures Improve Risky Long-Term Health Decisions," *Journal of Applied Behavior Analysis* 49: 148–169.

Kirby, K.N. and Petry, N.M. (2004). "Heroin and Cocaine Abusers Have Higher Discount Rates for Delayed Rewards than Alcoholics or Non-Drug-Using Controls," *Addiction* 99: 461–471.

Koffarnus, M.N. and Bickel, W.K. (2014). "A 5-trial Adjusting Delay Discounting Task: Accurate Discount Rates in Less than One Minute," *Experimental and Clinical Psychopharmacology* 22: 222–228.

Koffarnus, M.N., Franck, C.T., Stein, J.S., and Bickel, W.K. (2015). "A Modified Exponential Behavioral Economic Demand Model to Better Describe Consumption Data," *Experimental and Clinical Psychopharmacology* 23: 504–512.

Koffarnus, M.N., Jarmolowicz, D.P., Mueller, E.T., and Bickel, W.K. (2013). "Changing Delay Discounting in the Light of the Competing Neurobehavioral Decision Systems Theory: A Review," *Journal of the Experimental Analysis of Behavior* 99: 32–57.

Kollins, S.H. (2003). "Delay Discounting Is Associated with Substance Use in College Students," *Addictive Behaviors* 28: 1167–1173.

Krishnan-Sarin, S., Reynolds, B., Duhig, A.M., Smith, A., Liss, T., McFetridge, A., et al. (2007). "Behavioral Impulsivity Predicts Treatment Outcome in a Smoking Cessation Program for Adolescent Smokers," *Drug and Alcohol Dependence* 88: 79–82.

MacKillop, J. (2016). "The Behavioral Economics and Neuroeconomics of Alcohol Use Disorders," *Alcoholism: Clinical and Experimental Research* 40: 672–685.

MacKillop, J. and Kahler, C.W. (2009). "Delayed Reward Discounting Predicts Treatment Response for Heavy Drinkers Receiving Smoking Cessation Treatment," *Drug and Alcohol Dependence* 104: 197–203.

MacKillop, J. and Munafo, M. (2013). "An Intermediate Phenotype Approach to Addiction Genetics," in J. MacKillop and M. Munafo (eds), *Genetic Influences on Addiction: An Intermediate Phenotype Approach* (pp. 1–19). Cambridge: MIT Press.

MacKillop, J. and Murphy, J.G. (2007). "A Behavioral Economic Measure of Demand for Alcohol Predicts Brief Intervention Outcomes," *Drug and Alcohol Dependence* 89: 227–233.

MacKillop, J., Amlung, M., Acker, J., Gray, J.C., Brown, C.L., Murphy, J.G., et al. (2014). "The Neuroeconomics of Alcohol Demand: An Initial Investigation of the Neural Correlates of Alcohol Cost-Benefit Decision Making in Heavy Drinking Men," *Neuropsychopharmacology* 39: 1988–1995.

MacKillop, J., Amlung, M.T., Few, L.R., Ray, L.A., Sweet, L.H., and Munafò, M.R. (2011). "Delayed Reward Discounting and Addictive Behavior: A Meta-Analysis," *Psychopharmacology (Berl)* 216: 305–321.

MacKillop, J., Brown, C.L., Stojek, M.K., Murphy, C.M., Sweet, L., and Niaura, R.S. (2012). "Behavioral Economic Analysis of Withdrawal- and Cue-Elicited Craving for Tobacco: An Initial Investigation," *Nicotine & Tobacco Research* 14: 1426–1434.

MacKillop, J., Miranda, R., Monti, P.M., Ray, L.A., Murphy, J.G., Rohsenow, D.J., et al. (2010). "Alcohol Demand, Delayed Reward Discounting, and Craving in Relation to Drinking and Alcohol Use Disorders," *Journal of Abnormal Psychology* 119: 106–114.

MacKillop, J., Murphy, C.M., Martin, R.A., Stojek, M., Tidey, J.W., Colby, S.M., et al. (2015). "Predictive Validity of a Cigarette Purchase Task in a Randomized Controlled Trial of Contingent Vouchers for Smoking in Individuals with Substance Use Disorders," *Nicotine & Tobacco Research* 18(5): 531-537.

MacKillop, J., Murphy, J.G., Ray, L.A., Eisenberg, D.T.A., Lisman, S.A., Lum, J.K., and Wilson, D.S. (2008). "Further Validation of a Cigarette Purchase Task for Assessing the Relative Reinforcing Efficacy of Nicotine in College Smokers," *Experimental and Clinical Psychopharmacology* 16: 57–65.

MacKillop, J., O'Hagen, S., Lisman, S.A., Murphy, J.G., Ray, L.A., Tidey, J.W., et al. (2010). "Behavioral Economic Analysis of Cue-Elicited Craving for Alcohol," *Addiction* 105: 1599–1607.

Madden, G.J. and Bickel, W.K. (2009). *Impulsivity: The Behavioral and Neurological Science of Discounting*. Washington, DC: American Psychological Association.

Madden, G.J., Begotka, A.M., Raiff, B.R., and Kastern, L.L. (2003). "Delay Discounting of Real and Hypothetical Rewards," *Experimental & Clinical Psychopharmacology* 11: 139–145.

Madden, G.J., Bickel, W.K., and Jacobs, E.A. (1999). "Discounting of Delayed Rewards in Opioid-Dependent Outpatients: Exponential or Hyperbolic Discounting Functions?" *Experimental and Clinical Psychopharmacology* 7: 284–293.

Manchikanti, L., Helm, S., Fellows, B., Janata, J.W., Pampati, V., Grider, J.S., et al. (2012). "Opioid Epidemic in the United States," *Pain Physician* 15: ES9–38.

Marshall, A. (1898). *Principles of Economics*. Vol. 1. London: Macmillan & Co.

Mazur, J.E. (1987). "An Adjusting Procedure for Studying Delayed Reinforcement," in J.E. Mazur, J.A. Nevin, and H. Rachlin (eds), *Quantitative Analysis of Behavior: Vol 5. The Effects of Delay and Intervening Events on Reinforcement Value* (pp. 55–73). Hillsdale: Erlbaum.

Mejía-Cruz, D., Green, L., Myerson, J., Morales-Chainé, S., and Nieto, J. (2016). "Delay and Probability Discounting by Drug-dependent Cocaine and Marijuana Users," *Psychopharmacology* 233(14): 2705–2714.

Mello, N.K. and Mendelson, J.H. (1965). "Operant Analysis of Drinking Patterns of Chronic Alcoholics," *Nature* 206: 43–46.

Miller, W.R., Walters, S.T., and Bennett, M.E. (2001). "How Effective Is Alcoholism Treatment in the United States?" *Journal of Studies on Alcohol* 62: 211–220.

Mitchell, S.H. (1999). "Measures of Impulsivity in Cigarette Smokers and Non-Smokers," *Psychopharmacology (Berl)* 146: 455–464.

Mokdad, A.H., Marks, J.S., Stroup, D.F., and Grerberding, J.L. (2004). "Actual Causes of Death in the United States," *Journal of the American Medical Association* 291: 1238–1245.

Monterosso, J., Ainslie, G., Xu, J., Cordova, X., Domier, C.P., and London, E.D. (2007). "Frontoparietal Cortical Activity of Methamphetamine-Dependent and Comparison Subjects Performing a Delay Discounting Task," *Human Brain Mapping* 28: 383–393.

Moyer, A. and Finney, J.W. (2002). "Outcomes for Untreated Individuals Involved in Randomized Trials of Alcohol Treatment," *Journal of Substance Abuse Treatment* 23: 247–252.

Murphy, J.G. and MacKillop, J. (2006). "Relative Reinforcing Efficacy of Alcohol Among College Student Drinkers," *Experimental and Clinical Psychopharmacology* 14: 219–227.

Murphy, J.G., Correia, C.J., and Barnett, N.P. (2007). "Behavioral Economic Approaches to Reduce College Student Drinking," *Addictive Behaviors* 32: 2573–2585.

Murphy, J.G., Dennhardt, A.A., Skidmore, J.R., Borsari, B., Barnett, N.P., Colby, S.M., et al. (2012). "A Randomized Controlled Trial of a Behavioral Economic Supplement to Brief Motivational Interventions for College Drinking," *Journal of Consulting and Clinical Psychology* 80: 876–886.

Murphy, J.G., Dennhardt, A.A., Yurasek, A.M., Skidmore, J.R., Martens, M.P., MacKillop, J., et al. (2015). "Behavioral Economic Predictors of Brief Alcohol Intervention Outcomes," *Journal of Consulting and Clinical Psychology* 83: 1033–1043.

Murphy, J.G., Skidmore, J.R., Dennhardt, A.A., Martens, M.P., Borsari, B., Barnett, N.P., et al. (2012). "A Behavioral Economic Supplement to Brief Motivational Interventions for College Drinking," *Addiction Research & Theory* 20: 456–465.

Myerson, J., Green, L., and Warusawitharana, M. (2001). "Area Under the Curve as a Measure of Discounting," *Journal of Experimental Analysis of Behavior* 76: 235–243.

National Drug Intelligence Center (2011). *National Drug Threat Assessment*.

National Institute on Alcohol Abuse and Alcoholism. (2005). "National Advisory Council on Alcohol Abuse and Alcoholism – Recommended Council Guidelines on Ethyl Alcohol Administration in Human Experimentation – Revised May 2005." Available at http://www.niaaa.nih.gov/Resources/ResearchResources/Pages/job22.aspx (accessed November 15, 2016).

National Institute on Drug Abuse (2012). "Contingency Management Interventions/ Motivational Incentives (Alcohol, Stimulants, Opioids, Marijuana, Nicotine)," in *Principles of Drug Addiction Treatment: A Research-Based Guide* (3rd edition).

Nelson, L.S. and Perrone, J. (2012). "Curbing the Opioid Epidemic in the United States: The Risk Evaluation and Mitigation Strategy (REMS)," *JAMA* 308: 457–458.

Odum, A.L., Madden, G.J., and Bickel, W.K. (2002). "Discounting of Delayed Health Gains and Losses by Current, Never- and Ex-Smokers of Cigarettes," *Nicotine & Tobacco Research* 4: 295–303.

Office of National Drug Control Policy (2004). *The Economic Costs of Drug Abuse in the United States*. Washington, DC.

Owens, M.M., Murphy, C.M., and MacKillop, J. (2015). "Initial Development of a Brief Behavioral Economic Assessment of Alcohol Demand," *Psychology of Consciousness* 2: 144–152.

Owens, M.M., Ray, L.A., and MacKillop, J. (2015). "Behavioral Economic Analysis of Stress Effects on Acute Motivation for Alcohol," *Journal of the Experimental Analysis of Behavior* 103: 77–86.

Peters, J. and Buchel, C. (2010). "Episodic Future Thinking Reduces Reward Delay Discounting Through an Enhancement of Prefrontal-Mediotemporal Interactions," *Neuron* 66: 138–148.

Petry, N.M. (2001). "Delay Discounting of Money and Alcohol in Actively Using Alcoholics, Currently Abstinent Alcoholics, and Controls," *Psychopharmacology (Berl)* 154: 243–250.

Petry, N.M., Alessi, S.M., and Ledgerwood, D.M. (2012). "A Randomized Trial of Contingency Management Delivered by Community Therapists," *Journal of Consulting and Clinical Psychology* 80: 286–298.

Petry, N.M., Martin, B., Cooney, J.L., and Kranzler, H.R. (2000). "Give Them Prizes and They Will Come: Contingency Management for Treatment of Alcohol Dependence," *Journal of Consulting and Clinical Psychology* 68: 250–257.

Popovici, I., Homer, J.F., Fang, H., and French, M.T. (2012). "Alcohol Use and Crime: Findings from a Longitudinal Sample of U.S. Adolescents and Young Adults," *Alcoholism, Clinical and Experimental Research* 36: 532–543.

Prendergast, M., Podus, D., Finney, J., Greenwell, L., and Roll, J. (2006). "Contingency Management for Treatment of Substance Use Disorders: A Meta-Analysis," *Addiction (Abingdon, England)* 101: 1546–1560.

Roozen, H.G., Boulogne, J.J., van Tulder, M.W., van den Brink, W., De Jong, C.A.J., and Kerkhof, A.J.F.M. (2004). "A Systematic Review of the Effectiveness of the Community Reinforcement Approach in Alcohol, Cocaine and Opioid Addiction," *Drug and Alcohol Dependence* 74: 1–13.

SAMHSA (2013). *2013 National Survey on Drug Use and Health (NSDUH)*.

Schumacher, J.E., Milby, J.B., Wallace, D., Meehan, D.C., Kertesz, S., Vuchinich, R., et al. (2007). "Meta-analysis of Day Treatment and Contingency-Management Dismantling Research: Birmingham Homeless Cocaine Studies (1990–2006)," *Journal of Consulting and Clinical Psychology* 75: 823–828.

Secades-Villa, R., Weidberg, S., García-Rodríguez, O., Fernández-Hermida, J.R., and Yoon, J.H. (2014). "Decreased Delay Discounting in Former Cigarette Smokers at One Year After Treatment," *Addictive Behaviors* 39: 1087–1093.

Sheffer, C., MacKillop, J., McGeary, J., Landes, R., Carter, L., Yi, R., et al. (2012). "Delay Discounting, Locus of Control, and Cognitive Impulsiveness Independently Predict Tobacco Dependence Treatment Outcomes in a Highly Dependent, Lower Socioeconomic Group of Smokers," *American Journal on Addictions* 21: 221–232.

Smith, J.E., Meyers, R.J., and Miller, W.R. (2001). "The Community Reinforcement Approach to the Treatment of Substance Use Disorders," *American Journal on Addictions* 10: 51–59.

Snider, S.E., LaConte, S.M., and Bickel, W.K. (2016). "Episodic Future Thinking: Expansion of the Temporal Window in Individuals with Alcohol Dependence," *Alcoholism: Clinical and Experimental Research*.

Stanger, C., Ryan, S.R., Fu, H., Landes, R.D., Jones, B.A, Bickel, W.K., et al. (2012). "Delay Discounting Predicts Adolescent Substance Abuse Treatment Outcome," *Experimental and Clinical Psychopharmacology* 20: 205–212.

Stitzer, M. and Petry, N.M. (2006). "Contingency Management for Treatment of Substance Abuse," *Annual Review of Clinical Psychology* 2: 411–434.

Stockwell, T., Auld, M.C., Zhao, J., and Martin, G. (2012). "Does Minimum Pricing Reduce Alcohol Consumption? The Experience of a Canadian Province," *Addiction* 107: 912–920.

Stockwell, T., Zhao, J., Marzell, M., Gruenewald, P.J., Macdonald, S., Ponicki, W.R., et al. (2015). "Relationships Between Minimum Alcohol Pricing and Crime During the Partial Privatization of a Canadian Government Alcohol Monopoly," *Journal of Studies on Alcohol and Drugs* 76: 628–34.

Treno, A.J., Marzell, M., Gruenewald, P.J., and Holder, H. (2014). "A Review of Alcohol and other Drug Control Policy Research," *Journal of Studies on Alcohol and Drugs* 75: 98–107.

Vuchinich, R.E. and Heather, N. (2003). *Choice, Behavioural Economics and Addiction*. Amsterdam: Elsevier.

Wagenaar, A.C., Salois, M.J., and Komro, K.A. (2009). "Effects of Beverage Alcohol Price and Tax Levels on Drinking: A Meta-analysis of 1003 Estimates from 112 Studies," *Addiction* 104: 179–190.

Wagenaar, A.C., Tobler, A.L., and Komro, K.A. (2010). "Effects of Alcohol Tax and Price Policies on Morbidity and Mortality: A Systematic Review," *American Journal of Public Health* 100: 2270–2278.

Walters, G.D. (2014). "Crime and Substance Misuse in Adjudicated Delinquent Youth: The Worst of Both Worlds," *Law and Human Behavior* 38: 139–150.

Washio, Y., Higgins, S.T., Heil, S.H., McKerchar, T.L., Badger, G.J., Skelly, J.M., et al. (2011). "Delay discounting is associated with treatment response among cocaine-dependent outpatients," *Experimental and Clinical Psychopharmacology* 19(3): 243.

Yoon, J.H., Higgins, S.T., Heil, S.H., Sugarbaker, R.J., Thomas, C.S., and Badger, G.J. (2007). "Delay Discounting Predicts Postpartum Relapse to Cigarette Smoking Among Pregnant Women," *Experimental and Clinical Psychopharmacology* 15: 176–186.

Yurasek, A.M., Dennhardt, A.A., and Murphy, J.G. (2015). "A Randomized Controlled Trial of a Behavioral Economic Intervention for Alcohol and Marijuana Use," *Experimental and Clinical Psychopharmacology* 23: 332–338.

5

BEHAVIORAL ECONOMICS: TOOLS FOR PROMOTION OF PHYSICAL ACTIVITY

Tammy Leonard and Kerem Shuval

Introduction: the importance of physical activity

Habitually engaging in physical activity, defined as movement that involves skeletal muscle contractions resulting in increased energy expenditure above resting rate (U.S. Department of Health and Human Services, 1996), has been linked to the prevention of numerous health conditions including cardiovascular diseases, type 2 diabetes, weight control, certain cancers (e.g., breast and colon), depression, recurrent falls, hip fractures, cognitive dysfunction, and premature death (U.S. Department of Health and Human Services, 1996, 2008). A causal relationship between physical activity and health is supported by ample evidence from randomized controlled trials, well-designed observational studies and basic science. For example, worldwide it has been estimated that insufficient physical activity results in a 16 percent increased risk for coronary heart disease, 20 percent risk for type 2 diabetes, 33 percent for breast cancer, 32 percent for colon cancer, and 28 percent for mortality from all-causes (Lee et al., 2012).

Public health physical activity guidelines

To avoid the increased health risks associated with physical inactivity, the U.S. Department of Health and Human Services recommends in the 2008 Physical Activity Guidelines that adults engage in 150 minutes of moderate intensity physical activity (e.g., brisk walking) and/or 75 minutes of vigorous intensity physical activity (e.g., swimming) per week. For children, 60 minutes of daily physical activity is recommended. Engaging in more physical activity will likely result in additional health benefits, based on studies finding a dose-response relationship between physical activity and morbidity and mortality prevention. For example, a cohort study that followed over 660,000 men and women for an average of

14.2 years, found a 31 percent lower risk for mortality among those engaging in physical activity at 1–2 times the recommended levels, whereas being physically active 3–4 times the recommended levels had a 37% reduced mortality risk (Arem et al., 2015).

Unfortunately, most individuals do not meet the recommended levels, let alone exceed them. According to nationally representative data in the US, only about half (49.2%) of adults meet physical activity guidelines, based upon self-report measures (Centers for Disease Control and Prevention, n.d.). The data based on objectively measured physical activity (i.e., using accelerometers) are even more dismal, indicating that less than 5 percent of US adults meet the recommended activity levels (Troiano et al., 2008). The rate of physical activity differs markedly by age and gender, with men more active than women, and children more active than adolescents and adults. Moreover, when examining accelerometer-determined physical activity by race/ethnicity among adults aged 20–59 years, there are differences, however, mostly in men. Specifically, among men, non-Hispanic whites on average engaged in 34.6 minutes a day of moderate to vigorous intensity physical activity (MVPA), non-Hispanic blacks engaged in an average of 37.9 minutes of MVPA daily, whereas Hispanics engaged in 45.7 minutes of MVPA on average (Troiano et al., 2008). In women, non-Hispanic whites' engagement averaged 19.7 minutes per day of MVPA, non-Hispanic blacks averaged 20.0 minutes of MVPA daily, and Hispanics averaged 22.1 minutes of MVPA (Troiano et al., 2008). Additionally, Troiano et al. (2008) found that 42 percent of children aged 6–11 years met physical activity guidelines, as determined by accelerometers, whereas only 8 percent of adolescents aged 12–15 years and 7.6 percent of 16–19-year-olds met these guidelines.

Assessment and trends in physical activity levels

Physical activity rates have declined over the past several decades due, in large part, to increased automation and the time-saving devices at home and on the job (Ng and Popkin, 2012; Shuval et al., 2015). Occupational related activity has declined as well, as has active transportation (e.g., cycling or walking in lieu of driving), and household work (Archer et al., 2013; Ng and Popkin, 2012). Recreational physical activity, however, has remained fairly constant (Archer et al., 2013; Ng and Popkin, 2012). The energy costs of physical activity are expressed as units of metabolic equivalent of task (MET). A MET of 1 reflects a resting state while sitting; a MET of 3–5.9 represents moderate intensity activity (e.g., mowing the lawn); and a MET≥6 indicates vigorous intensity activity (e.g., swimming laps) (Ainsworth et al., 2011). To compute total energy expenditure during the week, the MET value from the various types of physical activity is multiplied by the duration and frequency of activity. For example, an individual walking for pleasure (MET=3.5), three times a week for an hour each time, will result in a total of 630 MET minutes per week (i.e., $3.5_{\text{METs}} \star 180_{\text{minutes/week}}$).

Total physical activity in the US averaged 265 MET minutes per week for adults in 1965, while only 160 MET minutes per week in 2009. Further, average weekly MET minutes were estimated to be 126 METs by 2030 (Ng and Popkin, 2012). The Physical Activity Guidelines reported above correspond to 500 MET minutes weekly; thus both past and current levels of activity are alarmingly low. In addition, nationally representative US time-use data found that women allocated ~26 hours each week to household work in 1965 compared to ~13 hours weekly in 2010, with non-employed women doing less household work than employed women (Archer et al., 2013). Conversely, screen based media use (e.g., TV viewing) doubled during the study period from ~8 hours per week in 1965 to ~16 hours weekly in 2010 (Archer et al., 2013).

Costs of inadequate physical activity levels

Low levels of physical activity impair people's health and directly impact medical costs. A study by Katzmarzyk et al. (2000) estimated $2.1 billion Canadian dollars were spent on medical care related to physical inactivity. They estimated that boosting bodily activity by 10 percent would reduce medical expenditures by $150 million Canadian dollars per year. An examination of medical claims data from Blue Cross and self-reported physical activity data from the Behavioral Risk Factor Surveillance System estimated that the insurers' expenditures resulting from insufficient physical activity were $83.6 million US dollars, or $56 US dollars per person (Garrett et al., 2004). Additionally, individuals with high levels of cardiorespiratory fitness at midlife, a result of habitual physical activity and genetics, had reduced medical care costs at the end of life. The average annual cost of health care use was $7,569 US dollars among participants with high fitness levels, in comparison to $9,379 for patients who were moderately fit, and $12,811 for low fit patients. Similarly, average annual costs from cardiovascular disease alone were lowest among the highly fit group, as were the number of office visits to the physician (Bachmann et al., 2015).

Economists have shown that increased physical activity is associated with higher personal earnings but the direction of causality is not entirely clear (Lechner and Sari, 2015; Maruyama and Yin, 2012). Low levels of physical activity are associated with both lower income and higher health care costs, making individuals living in low-income neighborhoods particularly vulnerable to the negative consequences of physical inactivity. Low-income households frequently live in neighborhoods least amenable to physical activity (e.g., high crime, inadequate parks, unkempt sidewalks), lack transportation and/or wealth resources to circumvent these challenges by, for example, paying for gym memberships, and are least likely to have adequate health insurance. Booth and Hawley (2015), emphasize that to avoid the "economic death march" associated with physical inactivity and high health care costs it is paramount to implement interventions to markedly reduce the epidemic of physical inactivity.

Overview of the literature

Public health strategies to improve physical activity

Public health interventions attempting to increase physical activity have primarily focused on individual behavior change (Healy et al., 2013; Sallis et al., 2006). These interventions, however, are often not sustainable and cannot be disseminated to the community at large (Sallis et al., 2006). For example, the seminal Diabetes Prevention Program (DPP) study demonstrated that an intensive physical activity and dietary intervention succeeded more than medication (metformin) or a placebo in reducing the incidence of type 2 diabetes among pre-diabetics (Diabetes Prevention Program Research Group, 2002). Although the DPP lifestyle intervention proved to be a cost-effective approach over a 10-year period, some DPP programs have encountered difficulties in implementation in "real world" settings (Faridi et al., 2010; The Diabetes Prevention Program Research, 2012).

In contrast to individual-level interventions, an ecological approach acknowledges that physical activity is impacted by a multitude of factors on the individual (e.g., self-efficacy), social/cultural (peer-support), organizational (e.g., workplace), environmental (e.g., parks and trails), and policy (e.g., school sitting/standing policies) levels (Owen et al., 2011; Sallis et al., 2012; Sallis et al., 2006). Interventions on multiple levels might be more sustainable and effective than targeting the individual level alone. For example, a population level intervention (e.g., community rail trail) promoting walking and cycling was a cost-effective means of increasing the physical activity of community members (Laine and Kuvaja-ko, 2014). Additionally, interventions promoting active transport among children and adolescents through the community, schools and parents have been successful in facilitating behavior change (Chillón et al., 2011; Felipe et al., 2015). Land use policies can also enable physical activity, such as ensuring the connectivity of sidewalks and making the environment more welcoming. Heath et al. (2012) suggest that community wide environmental and policy changes be implemented to ease access to pertinent facilities, as well as improvements in urban design.

Economic strategies to increase physical activity

Traditionally, economists have focused on understanding how individuals make decisions related to physical activity and how these decisions can be explained by individuals' allocation of scarce time and financial resources. This decision framework was first applied to health in the Grossman model of health investment (Grossman, 1972, 2000). The Grossman model incorporates the idea that individuals are both consumers and producers of health, and has been extended to include physical activity as a critical component of health production (Cawley, 2004; Meltzer and Jena, 2010).

Economists have long posited that, like many other decisions, the decision to engage in health promoting physical activity is a rational response to the costs and benefits associated with physical activity. Individuals value health and physical activity, but they must allocate time and money across a wide array of competing activities and goods that are also valuable. Thus, an avenue for intervening to increase physical activity lies in increasing the incentives for physical activity. Numerous interventions have been developed that financially incentivize physical activity. Overall, these studies find that financial incentives hold promise for increasing physical activity in the short-run, but may be less successful at long-term change (Finkelstein et al., 2013; Finkelstein et al., 2008; Kane et al., 2004; Mitchell et al., 2013). For example, incentivized gym membership increased gym visits during the 1-month study window, but these visits remained elevated only for participants who had never used the gym prior to the study (Charles and Gneezy, 2009). The success or failure of financial incentives may be better understood by incorporating behavioral economics constructs. For example, financial incentives may fail if they signal that physical activity is something so onerous that one should be paid to do it (Gneezy et al., 2011). Additionally, group-based incentives have been found to be more effective than individual incentives (Kullgren et al., 2013). Importantly, the Grossman model and subsequent studies have assumed that individuals rationally consider all competing costs and benefits when deciding to engage in physical activity. However, behavioral economists have provided evidence suggesting that individuals do not always behave as predicted by the rational choice model (Kagel and Roth, 1995).

To this end, Grossman's model of health production has been expanded to include many of these behavioral economics constructs. These constructs have been empirically related to physical activity (Chiteji, 2010; Huston and Finke, 2003; McKillop and Leonard, 2013), and are now considered as prime levers for increasing physical activity alongside the implementation of more traditional public health and economic incentive approaches.

Behavioral economics approaches

Thus far, strategies for increasing physical activity across all disciplines have met with mixed results. In some cases, robust strategies have been developed but implementation is a challenge (Faridi et al., 2010) or long-term behavior change is not achieved (i.e., economic incentive programs) (Charles and Gneezy, 2009; Kane et al., 2004; Mitchell et al., 2013). In other cases, strategies meet with varied results in different study populations or contexts (Joseph et al., 2014). The remainder of this chapter will explore behavioral economics constructs and their potential for unifying public health and economic approaches.

Behavioral economics can be utilized to better understand the often-complex decision to engage in physical activity, which in turn will provide suggestions for improving implementation of established interventions and indicate why some interventions may be more effective in different contexts. We also offer some suggestions for future research utilizing cross-disciplinary strategies. Specifically, we address measurement and data collection considerations that will improve the study of behavioral economics constructs. These constructs may be important mediators of intervention effectiveness and better understanding of their role in physical activity interventions is needed to improve the incorporation of behavioral economics tools in randomized multi-level interventions.

Behavioral economics constructs related to physical activity

Behavioral economics provides a framework for understanding the decision to engage in health-promoting physical activity by focusing on the process by which individuals make judgments given limited information, and how they utilize these judgments to make choices (Camerer and Loewenstein, 2004). Deciding to engage in physical activity requires individuals to make judgments of costs and benefits in an uncertain environment. The environment is characterized by future and varied probabilistic benefits and more immediate costs that have a relatively higher probability of occurrence. Thus uncertainty occurs in two domains: "will the cost/benefit occur?" and "when will the cost benefit/ occur?". For example, one might go for an evening jog tonight in order to reduce the likelihood of cardiovascular disease 10 years from now. The certainty of the immediate costs of an evening jog (e.g., lost time, exhaustion) are salient, but one is uncertain whether cardiovascular disease will be a health concern 10 years from now, or the degree to which the evening jog tonight will be influential in ameliorating that risk. Despite the complexity of a fully rational judgment, individuals make these decisions daily. As a result, we observe judgments related to the uncertain nature of cost/benefits of physical activity that side-step the inherent complexities of a fully rational approach and instead rely upon oversimplification of the decision frame, seek cues to resolve the uncertainty, or ignore uncertainty completely by simply basing behaviors on past habits (Camerer and Loewenstein, 2004).

Observed choices regarding the level of an individual's physical activity are a result of both judgments (which thereby become the information upon which the decision is based) and an individual's preferences—their relative like or dislike (e.g., for physical exhaustion, or lower risk of cardiovascular disease). In addition to assuming that individuals make fully rational judgments, the standard economic model (such as the Grossman model) also implicitly assumes that preferences are stable, invariant, and not context dependent. However, upon closer inspection, behavioral economists have found that preferences are not stable (Slovic, 1995). Instead, individuals "construct preferences" based upon the context at hand (Payne et al., 2003). Preferences are malleable. Thus in addition to a need to make frequent choices based upon uncertain judgments, we observe people making choices pertaining to physical activity that vary with the decision context and with time.

Behavioral regularities related to resolving uncertainty and intertemporal decisions

The decision to engage in physical activity has two key characteristics addressed by the behavioral economics approach: (1) individuals must make judgments related to risk and uncertainty and (2) physical activity decisions involve intertemporal choices, or choices that have costs and benefits that occur at different points in time.

Risk and uncertainty

Individuals tend to deviate from the standard economic model regarding the resolution of uncertainty and risk when the decision involves events that occur with very high or very low probabilities (Camerer and Loewenstein, 2004). Physical activity, as a preventive health behavior, might be directed toward the prevention of low-probability events. For example the 10-year risk of dying from a heart attack is less than 1 percent for a healthy young adult.[1] As a result, individuals might deviate from the prediction of the standard economic model in several ways.

First, individuals experience loss aversion because they evaluate losses and gains differently (Tversky and Kahneman, 1991). A loss and a gain of equivalent magnitudes are viewed differently; the loss is weighted more heavily. Thus the lost time and energy associated with physical activity is both more certain and will tend to be more heavily weighted than the gain of improved health.

Second, individual judgments are influenced by the current set of circumstances. This tendency takes many forms. Judgments are reference dependent in that they vary depending upon the way in which the uncertainty is presented or framed (Kahneman and Tversky, 1979; Zimmerman, 2009). Judgments are also frequently made out of habit without consideration (or re-consideration) of the information at hand, giving rise to status quo bias and anchoring (Roberto and Kawachi, 2014; Samuelson and Zeckhauser, 1988). For example, a young adult who is healthy and active might be going for five-kilometer runs daily, not because they consider the cost/benefits of the decision each and every day, but simply because it's "what they do" routinely. In addition, anchoring is a tendency to base a decision on the initial information provided without considering information that may later become evident or require some additional search. Specifically, one's own behavior in the past as well as present surrounding social norms can serve as an anchor, which, in turn, can impact behavior in general and physical activity in particular (Zimmerman, 2009). Habit formation of an active lifestyle during early childhood can serve as an anchor for sustained physical activity later on. Additionally, anchoring may explain why social pressures are so effective at influencing physical activity levels, although the most influential peers have proven to be those with the lowest levels of physical activity (Carrell et al., 2011; John and Norton, 2013; Kullgren et al., 2014).

Intertemporal choices

In order to engage in physical activity, individuals often have to reconcile how they *want* to use their leisure time for other purposes (e.g., screen time, sleep) with the way they know that they *should* use that time (e.g., exercising) to improve their health in the future. This "want/should" tradeoff involves intertemporal choices or decisions with costs and benefits that occur at different points in time (Milkman et al., 2008). In studies of savings behavior, people value money in the future significantly less than they value money today, that is, one dollar today has been viewed as equivalent to $0.53 one year from today (Angeletos et al., 2001). It is not surprising that savings rates are generally sub-optimal since even the most rewarding investment options fail to provide a comparably high level of return. Because the costs of physical activity occur in the present while the benefits occur in the future, individuals are likely to discount the future health benefits associated with physical activity. In fact, empirical evidence exists linking savings behavior, self-efficacy, future planning, and other measures of impatience with physical activity levels (Schwarzer et al., 2007; Shuval et al., 2014).

Individuals also tend to take a portfolio approach to thinking about intertemporal decisions balancing "good" and "bad" decisions over time (Milkman et al., 2008). Some studies refer to this as "compensatory behavior." For example, individuals in an intervention aimed at increasing physical activity, were found to have significantly higher levels of physical activity (suggesting an effective intervention), but no difference in body composition (Rosenkilde et al., 2012). The study authors suspect that participants engaged in offsetting behaviors in domains not impacted by the intervention, such as increased caloric intake or being more sedentary outside of the measured physical activity sessions. Finally, decision-making over intertemporal choices appears to be even more sub-optimal when it occurs alongside other cognitively challenging situations (Shiv and Fedorikhin, 1999). This is particularly problematic for want/should decisions regarding how to spend one's time—such as the decision to engage in physical activity. Cognitive loads are often the highest when time is most scarce. Thus sub-optimal decision-making is most likely to occur precisely when the need to optimize one's behavior is greatest.

Behavioral economics tools

One application of behavioral economics has been the development of tools that may be employed to guide the design of policy, interventions, and environments so as to make the "right" decision easier while not limiting the choice set, also known as asymmetric paternalism (Loewenstein et al., 2007; Wisdom et al., 2010). These tools are motivated by an understanding of how time and risk preferences can sometimes present challenges to decision-making and generally aid in overcoming one or more of the aforementioned behavioral regularities that result when individuals are faced with complex intertemporal choices over uncertain outcomes, such as the decision to engage in physical activity. A key aspect of behavioral economics tools for improving decisions is that they do not inhibit the individual's ability to choose, yet will make the "right" choice easier.

Public health interventions designed to increase physical activity often take a paternalistic approach by providing a "prescribed" amount of physical activity that participants in a program are asked to adhere to. While participants may choose to not follow the prescribed regime, such a prescription often has the effect of making engaging in physical activity less of a decision and more of an edict. Such interventions, while successful in increasing physical activity levels during the intervention, have limited long-term impact once the admonition is lifted and the perception of free choice is once again obtained (Epstein, 1998). Behavioral economics tools are unique in that they have the potential to insinuate a "prescription" for physical activity without actually being viewed as an edict; they aim to encourage the choice that involves physical activity while discouraging the inactive alternative (Thaler and Sunstein, 2008).

Defaults and environmental structures

An important tool for assisting individuals in optimally assessing risk and uncertainty is careful construction of the default behavior from which all other choices are evaluated. For example, workplace culture can establish default behaviors that are for or against engaging in physical activity during the workday. Stairwells can be inviting, welcoming wide places with background music, or desolate, poorly lit areas. Point of decision prompts can be placed in prominent places to encourage making the "right" decision, such as taking the stairs instead of the elevator (Soler et al., 2010). In addition, industry campuses can be designed to promote walking between office buildings with parking and roads only along the periphery or they can place buildings so that large parking lots and roads separate them. Companies can support walking clubs, work breaks, and provide "active workstations" (e.g., treadmill desks). Neighborhoods can be built with open green spaces, playgrounds, wide side-walks and bike trails to facilitate activity (Sallis et al., 2006). When neighborhoods are designed to encourage activity and more residents walk and bike, the shift in default behavior is made possible. For example, a study conducted in Curitiba, Brazil, observed that neighborhood walkability was positively associated with increased physical activity irrespective of income (Reis et al., 2013). Similarly, standing desks, active recess and physical education classes could be the default in school thereby decreasing sedentary behavior while increasing activity (Neuhaus et al., 2014).

Incentives

As mentioned previously, financial incentives have the potential to encourage physical activity by moving the benefits from the future into the present, thereby overcoming present-time bias. Behavioral economists have contributed to this work by hypothesizing why some incentive schemes might be more effective than others. For example, incentives for physical activity have shown some promise in increasing activity even after the incentives have ended because they change habits, or an individual's default behavior (Acland and Levy, 2010; Charles and Gneezy, 2009). A systematic review by Mitchell et al. (2013) examined the design attributes of studies where incentives successfully encouraged adherence to physical activity. The authors observed that incentivizing based on objectively measured exercise (e.g., via pedometers) rather than self-reporting was successful in producing meaningful behavior change. Additionally, guaranteed incentives (rather than lottery based), and larger incentives produced more favorable effects, though the sums of the incentives were relatively low (~\$3–~\$47). Further, more frequent incentives have the potential to bring the benefits of physical activity into the present, particularly pertinent for those with impatient time preferences. Nonetheless, additional research into the specific design needed to establish longer-term activity, and how individual time preferences are related to incentive effectiveness is clearly warranted.

Pre-commitment

One reason why intertemporal want/should decisions are often difficult is because our brain engages in different mental processes when considering immediate versus more distant decisions. Immediate decision-making is less deductive and information seeking, while future decisions are considered in greater detail (Milkman et al., 2008). Thus tools designed to aid in immediate decision-making should embrace simplicity, such as the "Let's Move" campaign aimed at increasing physical activity among youth, rather than offering elaborate explanations of benefits in meeting physical activity guidelines ("Let's Move!," n.d.; Roberto and Kawachi, 2014; Zimmerman, 2009). Additionally, pre-commitment devices are an effective tool of moving the decision frame from the immediate future to a more distant future when cognitive processes tend to be more information seeking and rational regarding weighing the costs and benefits of want/should decisions. For example, buying a parking decal that requires one to walk an extra 500 yards to and from work each day is one way in which individuals might pre-commit to engage in physical activity on a daily basis without having to make the decision repeatedly each morning on the drive into work when excuses for parking closer (i.e., the "want" as opposed to the more distant "should") are likely predominate.

Evidence from gym membership purchases and utilization behaviors suggest that individuals may be willing to engage in costly pre-commitment to address behavioral tendencies that often present hurdles to making physically active choices. Specifically, individuals are more likely to pay monthly gym membership fees rather than using pay-as-you go alternatives. However, on average, gym members would have saved significant amounts of money by using the pay-as-you-go option because they did not use the gym often enough to benefit from any pricing advantage inherent in a flat monthly fee (Epstein, 1998). Individuals are either over-confident about future gym use, or they are willing to pay a premium to pre-commit to gym use, even if it is at a lower rate than they had hoped for. It is difficult to differentiate between these two alternative explanations since we do not know whether pre-commitment (by paying the monthly fee) causes increased gym use over what would have occurred under a pay-as-you-go plan (Epstein, 1998).

Framing

Framing is a tool that may be used to address decision-making challenges arising from both uncertainty and intertemporal choices. Physical activity may be framed in ways that make the costs or benefits more salient (Zimmerman, 2009). A meta-analysis of 94 studies of health intervention messaging found that framing around the gains of preventive health behaviors as opposed to loss-based framing was significantly more likely to increase preventive health behavior uptake, and in particular increased physical activity (Gallagher and Updegraff, 2012). Framing is also a consideration in creating community-wide strategies for improving physical activity. Framing physical activity as a common goal for which all actors play a positive (rather than negative) role has been suggested as a more effective strategy than attributing barriers to segments of the community. For example, the VERB™ campaign conducted by the Centers for Disease Control and Prevention in collaboration with marketing experts, applied a branding strategy to specifically increase exercise among tweens (Asbury et al., 2008). The campaign focused on creating a brand ("VERB™ It's what you do") that is a unique and fun experience for all tweens rather than pointing out lack of physical activity as a common problem.

Next steps in utilizing behavioral economics to increase physical activity

Behavioral economics tools hold promise for improving physical activity. However, the tools have rarely been incorporated into randomized intervention studies; and when incorporated, studies are often not designed to measure the direct impact of the behavioral tool alone. Thus conclusions pertaining to the efficacy and later on the effectiveness of behavioral economics tools for increasing physical activity are premature. The existing "evidence base" relies heavily on extrapolations from observational studies showing that behavioral economics constructs (e.g., time and risk preferences) are related to physical activity levels (Anderson and Mellor, 2008; Leonard et al., 2013; Shuval et al., 2013) or assertions that behavioral economics constructs inherent in some intervention designs are important to promote physical activity (Gallagher and Updegraff, 2012). Therefore, a more regimented, interdisciplinary research approach is necessary to improve our understanding of how behavioral economics tools may be utilized to improve the adoption and maintenance of physical activity using well designed randomized controlled trials.

Measurement

In order to improve the reliability of randomized interventions designed to test the use of behavioral economics tools for increasing physical activity levels, rigorous standards of measurement for both physical activity and behavioral economics constructs must be employed. Studies that employ rigorous measurement in one domain often fall short in the other, frequently due to funding limitations, or lack of cross-disciplinary knowledge. Highly integrated multidisciplinary research teams can contribute by helping to establish guidelines for measurement along with discussion of how best to measure the constructs, given limited budgets. These guidelines can aide researchers in developing new studies or data collection efforts, and will also help to standardize the field so that studies are more comparable.

Measurement of physical activity (self-report "versus" objective)

Much of the population-wide evidence on the health benefits of habitually engaging in physical activity stems from self-reported physical activity and its impact on morbidity and mortality outcomes (Haskell, 2012; U.S. Department of Health and Human Services, 1996, 2008). Large cohort studies, such as Harvard's Nurses' Health Study (Manson et al., 1999), rely on self-reported physical activity data in which participants report the type, frequency and duration of activity. Self-reported information is subject to recall-bias and is likely, in the case of physical activity, to be an overestimation of the actual behavior. Indeed, data gathered from accelerometers, a small device that assesses the intensity levels of physical activity via the measurement of accelerations, confirms that the physical activity levels are much lower than self-report (Shuval et al., 2015). However, it is important to note that while accelerometers are an objective measure of physical activity, they have inherent limitations and should not be considered the gold standard. Examples of limitations are the relatively brief measurement period (usually worn for a week) and they do not measure all types of activity (e.g., swimming, cycling). Moreover, accelerometers do not measure the location or context of the activity (Haskell, 2012), and wearing a device on the waist might result in the Hawthorne effect, that is, participants change their behavior because they are being observed (McCarney et al., 2007; Smith-Spangler et al., 2007).

Measurement of behavioral economics constructs

The physical activity decision is related to two key economic constructs often studied by behavioral economists to explain deviations from predictions of the rational choice model: risk preferences and time preferences. Risk preferences describe the degree to which individuals are willing to take on risk. In general, most people try to avoid risk, which gives rise to risk averse preferences. However, people vary in terms of the degree to which they are risk averse. Time preferences describe the relative value of a good at an earlier date compared to its valuation at a later date. Someone with patient time preferences will be more likely to save money, while someone with impatient time preferences would be unlikely to sacrifice enjoyment in the present for some tangible, but future reward.

The gold standard for measurement of these constructs is economic experiments (Davis and Holt, 1993; Kagel and Roth, 1995). However, these are costly data to collect and the context-free nature of economic experiments sometimes inhibits our ability to measure preferences that are relevant in situations when context matters. Additionally, while experimental economists have developed standards for the conduct of economic experiments and some general guidelines for how to measure time and risk preferences, there is lack of consensus on the "perfect" experimental design, particularly for time preferences. Researchers wishing to implement economic experiments to measure time and risk preferences should implement tested protocols that have been validated in sub-populations most similar to the population being studied. For risk preference elicitation, the most widely used protocol is from Holt and Laury (2002). There is less consensus regarding the best experimental method for eliciting time preferences, but the most widely used methods involve asking participants to choose between receiving a larger amount of money in the future versus a smaller amount of money today (Frederick et al., 2002).

When economic experiments are too costly to implement, then self-report proxies of time and risk preferences may be used. However, proxy measures often do not correlate well with experimental measures and/or explain outcomes as one would expect (Anderson and Mellor, 2009; Burks et al., 2012). Future work exploring survey-based measures is necessary to improve the reliability with which time and risk preference measures can be incorporated into surveys.

Incorporation of behavioral economics tools within behavioral interventions

The early part of this chapter discussed several behavioral economics tools that may be incorporated into multi-level behavioral interventions aimed at modifying physical activity levels. To date, the evidence base for when and how these tools work is lacking. It is likely that some tools may be more/less effective in different contexts, which include multiple levels and sectors. Behavioral economics tools might be particularly useful in overcoming existing challenges at the individual level. For example, pre-commitment devices may be helpful for increasing exercise levels. However, because we know that there are multiple levels that influence the decision to engage in physical activity, behavioral economics tools must be used within a multi-level perspective. A given tool might only be helpful when implemented beyond the individual level. For example, due to peer effects and the reported tendency of people to be influenced by others' lack of physical activity (John and Norton, 2013), pre-commitment devices to improve gym use may be most effective if enacted at the neighborhood level such that all gym users engage in pre-commitment. Nevertheless, these ideas remain speculative and warrant further study. It is important for future researchers to report carefully on the type of behavioral economics tool used and the context in which it is implemented so that study results can be compared to determine which tools are most effective, for which sub-populations, and in which contexts. Randomized multi-level interventions will be best equipped to provide compelling answers to these questions.

Behavioral economics constructs as modifiers

It is also quite possible that behavioral economics constructs (e.g., time and risk preferences) are important modifiers in physical activity behaviors. Correlations between time and risk preferences and physical activity as well as other preventive health behaviors have been widely reported. This suggests that time and risk preferences might also be important modifiers for the effectiveness of behavioral economics tools and intervention participation. However, no known studies have examined these possibilities. Future studies should measure time and risk preferences as an important aspect of the individual-level context in which interventions are implemented. A better understanding of time and risk preferences as modifiers will improve future intervention design and targeting. Further, this work may also inform a clearer understanding of the endogenous determinants of time and risk preferences for the development of new tools aimed at overcoming challenges that impatience and risk aversion pose in increasing physical activity.

Conclusion

The decision to engage in physical activity is challenging because it involves both uncertainty (regarding the future benefits) and intertemporal choices (giving up time/comfort now in exchange for improved health in the future). The way in which individuals handle both uncertainty and intertemporal choices is related to physical activity levels. Thus, there is likely much to be gained from leveraging behavioral economics tools designed to improve intertemporal decision-making in uncertain environments, and to design and target interventions for increasing both individual and population physical activity levels. Researchers and intervention designers have a wealth of established behavioral economics tools to choose from, but these tools have rarely been applied to the field of physical activity. Future randomized behavioral interventions incorporating behavioral economics tools and constructs have great potential for high impact advances in understanding how to improve physical activity levels.

The research agenda in this area should include key elements. First, there is a need for better dissemination of validated methods for measuring both physical activity and behavioral economics constructs to improve the quality of cross-disciplinary studies. Second, increased incorporation of these measures in future studies will help researchers elucidate the role of behavioral economics constructs as mediators of physical activity and/or intervention effectiveness. Third, implementation of multi-level randomized behavioral interventions is necessary to test the effectiveness of behavioral economics tools for increasing physical activity. Behavioral economists and public health researchers alike have much to gain from a joint research agenda that can improve both public and private efforts to increase physical activity, which in turn is a prime lever for reducing risk of morbidity from a host of chronic diseases.

Note

1 Computed using the Framingham Risk Score (http://cvdrisk.nhlbi.nih.gov/calculator.asp).

Assuming a 26-year-old female; total cholesterol 190 mg/dl; HDL cholesterol 45; Systolic blood pressure: 110 mm/Hg; Non Smoker; and not currently taking medication for high blood pressure.

References

Acland, D., and Levy, M. (2010). "Habit Formation and Naivete in Gym Attendance: Evidence from a Field Experiment," unpublished Paper.

Ainsworth, B.E., Haskell, W.L., Herrmann, S.D., Meckes, N., Bassett, D.R., Tudor-Locke, C., et al. (2011). "2011 Compendium of Physical Activities: A Second Update of Codes and MET Values," *Medicine and Science in Sports and Exercise* 43(8): 1575–1581.

Anderson, L.R. and Mellor, J.M. (2008). "Predicting Health Behaviors with an Experimental Measure of Risk Preference," *Journal of Health Economics* 27(5): 1260–1274.

Anderson, L.R. and Mellor, J.M. (2009). "Are Risk Preferences Stable? Comparing an Experimental Measure with a Validated Survey-Based Measure," *Journal of Risk and Uncertainty* 39(2): 137–160. doi.org/10.1007/s11166-009-9075-z.

Angeletos, G.-M., Laibson, D., Repetto, A., Tobacman, J., and Weinberg, S. (2001). "The Hyperbolic Consumption Model: Calibration, Simulation, and Empirical Evaluation," *Journal of Economic Perspectives* 47–68.

Archer, E., Shook, R.P., Thomas, D.M., Church, T.S., Katzmarzyk, P.T., Hébert, J.R., … Blair, S.N. (2013). "45-Year Trends in Women's Use of Time and Household Management Energy Expenditure," *PloS One* 8(2): e56620.

Arem, H., Moore, S.C., Patel, A., Hartge, P., de Gonzalez, A.B., Visvanathan, K., … Matthews, C. E. (2015). "Leisure Time Physical Activity and Mortality a Detailed Pooled Analysis of the Dose-Response Relationship," *JAMA : Internal Medicine* 20892(6): 959–967. doi.org/10.1001/jamainternmed.2015.0533.

Asbury, L.D., Wong, F.L., Price, S.M., and Nolin, M.J. (2008). "The VERB Campaign: Applying a Branding Strategy in Public Health," *American Journal of Preventive Medicine* 34(6 SUPPL.): 183–187. doi.org/10.1016/j.amepre.2008.03.010.

Bachmann, J.M., DeFina L.F., Franzini, L., Gao, A., Leonard, D.S., Cooper, K.H., Berry, J.D., and Willis, B.L. (2015). "Cardiorespiratory Fitness in Middle Age and Health Care Costs in Later Life," *Journal of the American College of Cardiology* 66(17): 1876–1885. doi.org/10.1016/j.jacc.2015.08.030.

Booth, F.W. and Hawley, J.A. (2015). "The Erosion of Physical Activity in Western Societies: An Economic Death March," *Diabetologia* 58(8): 1730–1734.

Burks, S., Carpenter, J., Götte, L., and Rustichini, A. (2012). "Which Measures of Time Preference Best Predict Outcomes: Evidence from a Large-Scale Field Experiment," *Journal of Economic Behavior & Organization* 84(1): 308–320. doi.org/10.1016/j.jebo.2012.03.012.

Camerer, C.F. and Loewenstein, G. (2004). "Behavioral Economics: Past, Present, Future," in C.F. Camerer, G. Loewenstein, and M. Rabin (eds), *Advances in Behavioral Economics* (pp. 3–51). New York: Princeton University Press.

Carrell, S.E., Hoekstra, M., and West, J.E. (2011). "Is Poor Fitness Contagious?" *Journal of Public Economics* 95(7–8): 657–663. doi.org/10.1016/j.jpubeco.2010.12.005.

Cawley, J. (2004). "An Economic Framework for Understanding Physical Activity and Eating Behaviors," *American Journal of Preventive Medicine* 27(0749-3797 (Print)): 117–125.

Centers for Disease Control and Prevention. (n.d.). "Exercise or Physical Activity," Available at www.cdc.gov/nchs/fastats/exercise.htm (accessed January 3, 2016).

Charles, G. and Gneezy, U. (2009). "Incentives to Exercise," *Econometrica* 77(3): 909–931. doi.org/10.3982/ECTA7416.

Chillón, P., Evenson, K.R., Vaughn, A., and Ward, D.S. (2011). "A Systematic Review of Interventions for Promoting Active Transportation to School," *International Journal of Behavioral Nutrition and Physical Activity* 8(10): doi: 10.1186/1479–5868–8–10.

Chiteji, N. (2010). "Time-preference, Non-cognitive Skills and Well-being Across the Life Course: Do Non-cognitive Skills Encourage Healthy Behavior?" *The American Economic Review* 100(2): 200–204. doi.org/10.1257/aer.100.2.200.Time-preference.

Davis, D.D. and Holt, C.A. (1993). *Experimental Economics*. Princeton, NJ: Princeton University Press.

Diabetes Prevention Program Research Group. (2002). "Reduction in the Incidence of Type 2 Diabetes with Lifestyle Intervention or Metformin," *New England Journal of Medicine* 346(6): 393–403.

Epstein, L.H. (1998). "Integrating Theoretical Approaches to Promote Physical Activity," *American Journal of Preventive Medicine* 15(4): 257–265.

Faridi, Z., Shuval, K., Njike, V.Y., Katz, J.A., Jennings, G., Williams, M., and Katz, D.L. (2010). "Partners Reducing Effects of Diabetes (PREDICT): A Diabetes Prevention Physical Activity and Dietary Intervention Through African-American Churches," *Health Education Research* 25(2): 306–315. doi.org/10.1093/her/cyp005.

Felipe, C., Gomez, L.F., Sarmiento, R., Fernanda, M., Reis, R., Quistberg, D.A., and Parra, D.C. (2015). "Urban Environment Interventions Linked to the Promotion of Physical Activity: A Mixed Methods Study Applied to the Urban Context of Latin America," *Social Science & Medicine* 131: 18–30. doi.org/10.1016/j.socscimed.2015.02.042.

Finkelstein, E.A., Brown, D.S., Brown, D.R., and Buchner, D.M. (2008). "A Randomized Study of Financial Incentives to Increase Physical Activity Among Sedentary Older Adults," *Preventive Medicine* 47(2): 182–187. doi.org/10.1016/j.ypmed.2008.05.002.

Finkelstein, E.A., Tan, Y.T., Malhotra, R., Lee, C.F., Goh, S.S., and Saw, S.M. (2013). "A Cluster Randomized Controlled Trial of an Incentive-Based Outdoor Physical Activity Program," *The Journal of Pediatrics* 163(1): 167–172.

Frederick, S., Loewenstein, G., and O'Donoghue, T. (2002). "Time Discounting and Time Preference: A Critical Review," *Journal of Economic Literature* 40(2): 351–401.

Gallagher, K.M. and Updegraff, J.A. (2012). "Health Message Framing Effects on Attitudes, Intentions, and Behavior: A Meta-Analytic Review," *Annals of Behavioral Medicine* 43(1): 101–116.

Garrett, N.A., Brasure, M., Schmitz, K.H., Schultz, M.M., and Huber, M.R. (2004). "Direct Cost to a Health Plan," *American Journal of Preventive Medicine* 27(4): 304–309. doi.org/10.1016/j.amepre.2004.07.014.

Gneezy, U., Meier, S., and Rey-Biel, P. (2011). "When and Why Incentives (Don't) Work to Modify Behavior," *Journal of Economic Perspectives* 25(4): 191–210. doi.org/10.1257/jep.25.4.191.

Grossman, M. (1972). "On the Concept of Health Capital and the Demand for Health," *The Journal of Political Economy* 80(2): 223–255.

Grossman, M. (2000). "The Human Capital Model," *Handbook of Health Economics* 1: 347–408.

Haskell, W.L. (2012). "Physical Activity by Self-Report: A Brief History and Future Issues," *Journal of Physical Activity and Health* 9(Suppl 1): 5–10.

Healy, G.N., Eakin, E.G., Lamontagne, A.D., Owen, N., Winkler, E.A., Wiesner, G., … Dunstan, D. W. (2013). "Reducing Sitting Time in Office Workers: Short-Term Efficacy of a Multicomponent Intervention," *Preventive Medicine* 57(1): 43–48.

Heath G.W., Parra D., Sarmiento, O.L., Anderson L.B., Owen N., Goenka S., Montes F., and Brownson R.C., for the Lancet Physician Activity Series Working Group (2012). "Evidence-based Intervention in Physical-Activity: Lessons from Around the World." *Lancet* 380: 282–291.

Holt, C.A. and Laury, S.K. (2002). "Risk Aversion and Incentive Effects," *American Economic Review* 92(5): 1644–1655.

Huston, S.J. and Finke, M.S. (2003). "Diet Choice and the Role of Time Preference," *The Journal of Consumer Affairs* 37(1): 143–160.

John, L.K. and Norton, M.I. (2013). "Converging to the Lowest Common Denominator in Physical Health," *Health Psychology* 32(9): 1023.

Joseph, R.P., Durant, N.H., Benitez, T.J., and Pekmezi, D.W. (2014). "Internet-based Physical Activity Interventions," *American Journal of Lifestyle Medicine* 8: 42–68.

Kagel, J.H. and Roth, A.E. (1995). *The Handbook of Experimental Economics. Economic Theory* (vol. 78). doi.org/10.2307/1243316.

Kahneman, D. and Tversky, A. (1979). "Prospect Theory: An Analysis of Decision Under Risk," *Econometrica* 47(2): 263–292. doi.org/10.1111/j.1536-7150.2011.00774.x.

Kane, R.L., Johnson, P.E., Town, R.J., and Butler, M. (2004). "A Structured Review of the Effect of Economic Incentives on Consumers' Preventive Behavior," *American Journal of Preventive Medicine* 27(4): 327–352. doi.org/10.1016/j.amepre.2004.07.002.

Katzmarzyk, P.T., Gledhill, N., and Shephard, R.J. (2000). "The Economic Burden of Physical Inactivity in Canada," *Canadian Medical Association Journal* 163(11): 1435–1440.

Kullgren, J.T., Harkins, K.A., Bellamy, S.L., Gonzales, A., Tao, Y., Zhu, J., … Karlawish, J. (2014). "A Mixed-Methods Randomized Controlled Trial of Financial Incentives and Peer Networks to Promote Walking Among Older Adults," *Health Education & Behavior : The Official Publication of the Society for Public Health Education* 41(1 Suppl): 43S–50S. doi.org/10.1177/1090198114540464.

Kullgren, J.T., Troxel, A.B., Loewenstein, G., Asch, D.A., Norton, L.A., Wesby, L., … Volpp, K.G. (2013). "Individual- Versus Group-Based Financial Incentives for Weight Loss: A Randomized, Controlled Trial," *Annals of Internal Medicine* 158(7): 505–514. doi.org/10.7326/0003-4819-158-7-201304020-00002.

Laine, J. and Kuvaja-ko, V. (2014). "Cost-Effectiveness of Population-Level Physical Activity Interventions: A Systematic Review," *American Journal of Health Promotion* 29(2): 71–81. doi.org/10.4278/ajhp.131210-LIT-622.

Lechner, M. and Sari, N. (2015). "Labor Market Effects of Sports and Exercise: Evidence from Canadian Panel Data," *Labour Economics* 35: 1–15. doi.org/10.1016/j.labeco.2015.04.001.

Lee, I.-M., Shiroma, E.J., Lobelo, F., Puska, P., Blair, S.N., and Katzmarzyk, P.T. (2012). "Effect of Physical Inactivity on Major Non-Communicable Diseases Worldwide: An Analysis of Burden of Disease and Life Expectancy," *Lancet* 380(9838): 219–229.

Leonard, T., Shuval, K., de Oliveira, A., Skinner, C.S., Eckel, C., and Murdoch, J.C. (2013). "Health Behavior and Behavioral Economics: Economic Preferences and Physical Activity Stages of Change in a Low-Income African-American Community," *American Journal of Health Promotion* 27(4): 211–221.

Let's Move! (n.d.). Available at http://www.bbc.co.uk/programmes/b03g64pk/episodes/downloads (accessed January 19, 2017).

Loewenstein, G., Brennan, T., and Volpp, K.G. (2007). "Asymmetric Paternalism to Improve Health Behaviors," *JAMA: The Journal of the American Medical Association* 298(20): 2415–2417.

Manson, J., Hu, F., Rich-Edwards, J., Colditz, G., Stampfer, M., Willett, W., ... Hennekens, C. (1999). "A Prospective Study of Walking as Compared with Vigorous Exercise in the Prevention of Coronary Heart Disease in Women," *New England Journal of Medicine* 341: 650–658.

Maruyama, S. and Yin, Q. (2012). "The Opportunity Cost of Exercise: Do Higher-Earning Australians Exercise Longer, Harder, or Both?" *Health Policy* 106(2): 187–194. doi.org/10.1016/j.healthpol.2012.03.013.

McCarney, R., Warner, J., Iliffe, S., van Haselen, R., Griffin, M., and Fisher, P. (2007). "The Hawthorne Effect: A Randomised, Controlled Trial," *BMC Medical Research Methodology* 7: 30.

McKillop, C. and Leonard, T. (2013). "The Impact of Individuals, Community and Neighborhood Context on Health Outcomes: Do Non-Cognitive Skills Such as Future Orientation and Social Cohesion Mediate Negative Influences?" Working Paper.

Meltzer, D.O. and Jena, A.B. (2010). "The Economics of Intense Exercise," *Journal of Health Economics* 29(3): 347–352. doi.org/10.1016/j.jhealeco.2010.03.005.

Milkman, K.L., Rogers, T., and Bazerman, M.H. (2008). "Harnessing Our Inner Angels and Demons: What We Have Learned About Want/Should Conflicts and How that Knowledge Can Help Us Reduce Short-Sighted Decision Making," *Perspectives on Psychological Science* 3(4): 324–338.

Mitchell, M.S., Goodman, J.M., Alter, D.A., John, L.K., Oh, P.I., Pakosh, M.T., and Faulkner, G.E. (2013). "Financial Incentives for Exercise Adherence in Adults: Systematic Review and Meta-Analysis," *American Journal of Preventive Medicine* 45(5): 658–667. doi.org/10.1016/j.amepre.2013.06.017.

Neuhaus, M., Eakin, E.G., Straker, L., Owen, N., Dunstan, D.W., Reid, N., and Healy, G.N. (2014). "Reducing Occupational Sedentary Time: A Systematic Review and Meta-Analysis of Evidence on Activity-Permissive Workstations," *Obesity Reviews : An Official Journal of the International Association for the Study of Obesity* 15(10): 822–838.

Ng, S.W. and Popkin, B.M. (2012). "Time Use and Physical Activity: A Shift Away From Movement Across The Globe," *Obesity Reviews: An Official Journal of the International Association for the Study of Obesity* 13(8): 659–680.

Owen, N., Sugiyama, T., Eakin, E.E., Gardiner, P.A., Tremblay, M.S., and Sallis, J.F. (2011). "Adults' Sedentary Behavior Determinants and Interventions," *American Journal of Preventive Medicine* 41(2): 189–196.

Payne, J.W., Bettman, J.R., and Johnson, E.J. (2003). "Behavioral Decision Research: A Constructive Processing Perspective," *Annual Review of Psychology*, 43: 87–132.

Reis, R.S., Akira, A., Hino, F., Rech, C.R., Kerr, J., and Hallal, P.C. (2013). "Walkability and Physical Activity: Findings from Curitiba, Brazil," *American Journal of Preventive Medicine* 45(3): 269–275. doi.org/10.1016/j.amepre.2013.04.020.Walkability.

Roberto, C.A. and Kawachi, I. (2014). "Use of Psychology and Behavioral Economics to Promote Healthy Eating," *American Journal of Preventive Medicine* 47(6): 832–837.

Rosenkilde, M., Auerbach, P., Reichkendler, M.H., Ploug, T., Stallknecht, B.M., and Sjödin, A. (2012). "Body Fat Loss and Compensatory Mechanisms in Response to Different Doses of Aerobic Exercise—A Randomized Controlled Trial in Overweight Sedentary Males," *American Journal of Physiology—Regulatory, Integrative and Comparative Physiology* 303(6): R571–R579.

Sallis, J.F., Cervero, R.B., Ascher, W., Henderson, K.A., Kraft, M.K., and Kerr, J. (2006). "An Ecological Approach to Creating Active Living Communities," *Annual Review of Public Health* 27, 297–322.

Sallis, J.F., Floyd, M.F., Rodríguez, D.A., and Saelens, B.E. (2012). "Role of Built Environments in Physical Activity, Obesity, and Cardiovascular Disease," *Circulation* 125(5): 729–737.

Samuelson, W. and Zeckhauser, R. (1988). "Status Quo Bias in Decision Making," *Journal of Risk and Uncertainty* 1(1): 7–59.

Schwarzer, R., Schüz, B., Ziegelmann, J.P., Lippke, S., Luszczynska, A., and Scholz, U. (2007). "Adoption and Maintenance of Four Health Behaviors: Theory-Guided Longitudinal Studies on Dental Flossing, Seat Belt Use, Dietary Behavior, and Physical Activity," *Annals of Behavioral Medicine* 33(2): 156–166.

Shiv, B. and Fedorikhin, A. (1999). "Heart and Mind in Conflict: The Interplay of Affect and Cognition in Consumer Decision Making," *Journal of Consumer Research* 26(3): 278–292.

Shuval, K., Barlow, C.E., Finley, C.E., Gabriel, K.P., Schmidt, M.D., and DeFina, L.F. (2015). "Standing, Obesity, and Metabolic Syndrome: Findings from the Cooper Center Longitudinal Study," *Mayo Clinic Proceedings*.

Shuval, K., Hébert, E.T., Siddiqi, Z., Leonard, T., Lee, S.C., Tiro, J.A, McCallister, B.A., and Skinner, C.S. (2013). "Impediments and Facilitators to Physical Activity and Perceptions of Sedentary Behavior Among Urban Community Residents: The Fair Park Study," *Preventing Chronic Disease* 10(1): E177.

Shuval, K., Nguyen, B.T., Yaroch, A.L., Drope, J., and Gabriel, K.P. (2015). "Accelerometer Determined Sedentary Behavior and Dietary Quality Among US Adults," *Preventive Medicine* 78, 38–43. doi.org/10.1016/j.ypmed.2015.06.010.

Shuval, K., Si, X., Nguyen, B., and Tammy, L. (2014). "Utilizing Behavioral Economics to Understand Adherence to Physical Activity Guidelines Among a Low-Income Urban Community," *Journal of Physical Activity and Health, Epub ahead.*

Slovic, P. (1995). "The Construction of Preference," *American Psychologist* 50: 364–371.

Smith-Spangler, C., Gienger, A.L., Lin, N., Lewis, R., Stave, C.D., and Olkin, I. (2007). "Using Pedometers to Increase Physical Activity: A Systematic Review," *JAMA: The Journal of the American Medical Association* 298: 2296–2304.

Soler, R.E., Leeks, K.D., Buchanan, L.R., Brownson, R.C., Heath, G.W., and Hopkins, D.H. (2010). "Point-of-Decision Prompts to Increase Stair Use: A Systematic Review Update," *American Journal of Preventive Medicine* 38(2): S292–S300.

Thaler, R.H. and Sunstein, C.R. (2008). *Nudge: Improving Decisions About Health, Wealth, And Happiness.* New Haven, CT: Yale University Press.

The Diabetes Prevention Program Research. (2012). "The 10-year Cost-Effectiveness of The Lifestlyle Intervention or Metformin for Diabetes Prevention," *Diabetes Care* 35, 723–730. doi.org/10.2337/dc11-1468.

Troiano, R.P., Berrigan, D., Dodd, K.W., Mâsse, L.C., Tilert, T., and McDowell, M. (2008). "Physical Activity in the United States Measured by Accelerometer," *Medicine and Science in Sports and Exercise* 40(1): 181–188.

Tversky, A. and Kahneman, D. (1991). "Loss Aversion in Riskless Choice: A Reference-Dependent Model," *The Quarterly Journal of Economics* 106(4): 1039–1061.

U.S. Department of Health and Human Services. (1996). *Physical Activity and Health: A Report of the Surgeon General.*

U.S. Department of Health and Human Services. (2008). *Physical Activity Guidelines for Americans.* Available at http://www.health.gov/paguidelines/pdf/paguide.pdf (accessed January 3, 2016).

Wisdom, B.J., Downs, J.S., and Loewenstein, G. (2010). "Promoting Healthy Choices: Information versus Convenience," *American Economic Journal: Applied Economics* 2(April): 164–178.

Zimmerman, F.J. (2009). "Using Behavioral Economics to Promote Physical Activity," *Preventive Medicine* 49(4): 289–291.

6

USING BEHAVIORAL ECONOMICS TO IMPROVE DIETARY INTAKE

Alternatives to regulation, bans, and taxation

Marie A. Bragg and Brian Elbel

Introduction

The role of the dietary intake in contributing to obesity

Interventions designed to improve dietary choices have historically focused on educational campaigns, the provision of information on nutrition, and programs focused on individual-level behavior change. Despite the proliferation of educational campaigns, commercial weight management programs, and clinical weight loss interventions, rates of overweight and obesity in the United States rose continuously during the past several decades and remain high. Indeed, commercial weight loss programs and clinical trials have shown limited weight loss with the high likelihood of regaining weight over time (Tsai and Wadden, 2005; Wadden et al., 2004; Young et al., 2012), demonstrating the immense behavioral change challenges faced by obese individuals and healthcare providers who counsel patients on weight loss. Over one-third of adults and one-fifth of children in the US are obese (Ogden et al., 2002, 2012, 2014) and those living in poverty face greater risk for becoming obese (Drewnowski and Specter, 2004; Malik et al., 2013).

Public health research focused on obesity prevention has recently shifted toward understanding how the various aspects of the environment contribute to the development of obesity and its comorbidities. This environment is made up of a variety of structural elements (e.g., public transportation; supermarket density), financial constraints (e.g., food prices), cultural norms (e.g., screen time; advertising), policies (e.g., taxes; wellness programs), and other factors that shape both physical activity and dietary behaviors. Indeed, studies show that consumer exercise habits and food choices are affected by the built environment (e.g., presence of sidewalks, safety of neighborhoods) (Booth et al., 2005; Sallis et al., 2012;

Mayne et al., 2015), availability of foods (e.g., proximity of supermarkets and fast-food restaurants) (Hoyt et al., 2014; Kruger et al., 2014), food and beverage advertisements (Parker et al., 2009), and a number of other factors that affect health choices (e.g., pricing and packaging of products) (Darmon et al., 2002). Within this environment, consumers make dozens of decisions every day regarding their food and beverage intake and physical activity (see Chapter 5 on Physical Activity), and those decisions shape long-lasting habits that affect health. Those decisions are determined, in part, by consumers' perceptions of the environment, biases and heuristics, and the way the environment is structured to encourage or deter certain health behaviors.

The food and beverage industry is a major player in shaping consumer perceptions, preferences, and behaviors. Food and beverage companies play a central role in determining how their products are priced, promoted through advertisements, and which types of products will be available for consumption. Portion sizes for sugar-sweetened beverages (SSBs), default side dish options served with meals at restaurants, and the placement of energy-dense, nutrient-poor products in supermarkets are just a few examples of how food and beverage companies shape the food environment. The current food environment heavily promotes highly processed foods, sugary beverages, and large portion sizes at relatively inexpensive prices, making it challenging for many consumers to maintain healthy dietary intake. Indeed, comparisons between portion weights and standard portions established by USDA and FDA show that cookies exceed USDA standards at 700 percent while cooked pasta, muffins, steaks, and bagels also exceed USDA standards at 480 percent, 333 percent, 224 percent, and 195 percent, respectively (Young and Nestle, 2002).

A variety of population-level, environmentally focused interventions have attempted to address obesity, but many of these efforts have been met with strong resistance from the food industry, policymakers, and consumers. Sugar-sweetened beverage taxes were proposed in a number of localities in the United States during the past couple of decades, but just one city (Berkeley, California) successfully passed a law (Mandaro, 2014). In Mexico, the SSB tax was associated with reductions in the purchases of taxed beverages and increases in the purchases of untaxed beverages (Colchero et al., 2016). In New York City, the proposed ban on transfat was linked to increased food costs and a decrease in food options by the food industry (Brownell and Pomeranz, 2014). Changes to school nutrition programs have improved the nutritional quality of meals served in educational institutions, despite persistent criticism and skepticism about the changes (French and Story, 2013; Woo Baidal and Taveras, 2014). Even menu labeling, which involves posting calorie information on chain restaurant menus, was met with strong opposition from the food industry (Tavernise, 2015). This policy, as multiple studies showed, was quite limited in its impact on food choice (Cantor et al., 2015; Bollinger et al., 2011). A key question is how an alternative approach to labeling might alter its effectiveness. Despite these promising approaches to obesity prevention, the difficulties in implementing these types of interventions suggest the need for supplementary approaches that appeal to consumers, the industry, and public health advocates.

Using economic approaches to address obesity

Some economic approaches hold promise as potential interventions for improving dietary choices. Traditional economic principles have been used to explain human behavior, but these approaches are limited by assumptions that people are rational, act according to their long-term self-interest, and that limited self-control does not impact the ability to behave as intended. Indeed, the traditional economic literature assumes that people make poor food choices because they lack perfect information that would help them act according to their best interest (Stigler, 1961). However, even in the presence of appropriate information, consumers are still affected by their biases, heuristics, and the way foods and beverages are presented.

Behavioral economics helps address some of the gaps of traditional economics by explaining the variety of nuances that influence decision-making; these nuances are especially relevant to food intake because of the competing factors involved in dietary choices (e.g., taste, convenience, price, health). Behavioral economics research demonstrates that consumers regularly shift from rational to impulsive decision-making based on the context of the consumer environment (Kahneman, 2003). These shifts in behavior can make healthy dietary behaviors difficult to maintain. Thaler and Sunstein (2009) propose one potential tool for changing behavior through "nudges," defined as "any aspect of the choice architecture that alters people's behavior in a predictable way without forbidding any options or significantly changing their economic incentives" while simultaneously being easy and cheap to avoid (p. 6). However, because nudges, by definition, are cheap and easy to avoid, one political limitation of behavioral economics involves the risk that "soft nudges" will be enacted instead of policies that restrict or eliminate a behavior and are unavoidable by consumers (e.g., SSB taxes, beverage portion cap policies, food advertising bans). Such changes to the food environment (the major focus of food policy advocates and policymakers) are perceived as more paternalistic than government regulation of the food environment. As such, these types of changes are more palatable to food and beverage companies and consumers who highly value personal responsibility and freedom of choice. However, behavioral economics approaches also need to be rigorously examined prior to implementation in order to increase the likelihood of effectiveness.

This chapter reviews behavioral economics concepts that can provide a useful framework for increasing healthy dietary decisions and shifting the food environment toward healthy defaults. Consumer judgments, valuations, and habits are shaped by factors such as biases and heuristics, default options, and preferences. Considering these factors will be critical for shifting the food environment away from heavily promoting energy-dense, nutrient-poor options and instead focus on purposeful presentation and promotion of products that are likely to encourage healthful dietary habits. Research shows that maintaining a healthy diet and consuming healthier foods can positively influence weight loss. For example, substitution of unhealthy refined-grain carbohydrates for healthier whole-grain carbohydrates can aid in reducing blood glucose levels and cardiovascular risk (Willett and Skerrett, 2001; Hu and Willett, 2002). Further, results from a meta-analysis show that obese adults lose more weight per week when eating a healthy diet or by coupling a healthy diet with exercise in comparison to just exercise alone (Miller et al., 1997). By highlighting the impact of dietary decisions on health, behavioral economics approaches to obesity prevention can strengthen existing policies and contribute to the development of new policies that preserve consumer choice while facilitating healthy behaviors.

Overview of the literature

Behavioral economics principles relevant to the study of eating behaviors

Traditional economic approaches generally consider that choices are shaped by preferences and budget constraints. In such a model, consumers make choices, including food choices, which are in their long-term best interest. They maximize their utility when faced with a budget constraint and a choice set of goods. Behavioral economics is an interdisciplinary study of psychology and economics that is rooted in the ideas that (1) most judgments and most choices are made intuitively and (2) that the rules that govern intuition are generally similar to the rules of perception (Kahneman, 2003). In terms of food behaviors, behavioral economics may also provide an additional explanation for obesity by evaluating risk from eating behaviors and time tradeoffs through individual preferences (Leonard, 2013).

Psychologists typically use System I and II thought processing to explain everyday behaviors. System I is automatic, effortless, and emotionally charged. System II is controlled, effortful, and neutral (Kahneman, 2003). One food environment, a supermarket for example, can affect food behaviors differently depending on the thought process used. Shopping while hungry may trigger System I thought processes and lead consumers to purchase more processed food items in order to curb hunger more quickly. A customer who is not influenced by hunger while grocery shopping is more likely to use System II thought processing, consider the lack of health benefits associated with consuming processed foods, and purchase healthier items such as fruits and vegetables. System II typically regulates System I through slower, logical thinking during information processing. However, depletion of cognitive resources by hunger, thirst, or stress allows System I's automatic, emotionally charged processing to prevail in decision-making (Kahneman, 2003). Through assessment of psychological and physical states (e.g., hunger) and consumer behavior, behavioral economics may be successful in explaining how thought processing can also influence eating behaviors.

A number of behavioral economics principles build upon the assumptions regarding Systems I and II, thus providing a variety of tools to consider when planning dietary behavior interventions. *Mass defaults* apply to all customers and services, without taking customers' individual preferences into account (Goldstein et al., 2008). For example, standard shipping is assigned by default regardless of when a customer needs an item (Goldstein et al., 2008). From a marketing perspective, food and beverage companies utilize larger product sizes in advertisements to draw more attention from all consumers, which has proven effective through an increase in profit. Restaurants use large dinner plates, bakers sell larger muffin tins, pizzerias use larger pans, and fast-food companies use larger drink and French fry containers (Young and Nestle, 2002).

Personalized defaults reflect individual differences and are better tailored to meet every customer's needs (Goldstein et al., 2008). For example, airlines automatically assign an aisle seat on a plane to individuals who have previously bought an aisle seat (Goldstein et al., 2008). Personalized defaults were also utilized in the "Share A Coke" campaign, where a wide variety of first names were printed on Coca-Cola cans and bottles. The ability to have one's name incorporated into the bottle and can design makes the product much more desirable to consume and to keep as a collector's item (Weiss, 2015).

Instant rewards can interfere with long-term decision-making. In terms of food behavior, the idea of eating or buying food now or later is best described by *immediate gratification*, *present-biased preferences* or *hyperbolic discounting* in behavioral economics terms (Frederick et al., 2002). When making food choices, we tend to pursue rewards that we do not appreciate in the long term (O'Donoghue and Rabin, 2000). For example, consider a meal from McDonald's consisting of a Big Mac, fries, and a large soda purchased during a lunch break from work. At the time, the food may taste good and satisfy hunger. However, after finishing the entire meal, consumers may feel sluggish and tired. Purchasing energy dense foods as opposed to preparing a healthy meal in advance might demonstrate present biased preferences if in making this choice the consumer did not fully consider longer-term self-interest. Although delayed in comparison to accessible energy dense foods, benefits of preparing a healthy lunch would include saving money and maintaining higher energy levels (e.g., productivity in the workplace).

Visceral factors are negative emotions, drive states, and feeling states that grab people's attention and motivate them to engage in specific behaviors (Loewenstein, 1996). The defining features of visceral factors include (1) direct hedonic impact, defined as the conscious and unconscious sensations that result from an experience (usually negative), and (2) an effect on the relative desirability of different goods and actions (Loewenstein, 1996). Imbalances between perceived self-interest and behavior result from visceral factors and can cause people to willingly behave in ways incongruent with their long-term goals and interests (Loewenstein, 1996). For example, someone who is thirsty will be motivated to quench this aversive feeling as quickly as possible. Visceral factors are often influenced by sensory cues such as sight, sound, or smell to trigger a craving. For example, the feeling of hunger makes food more desirable and predicts impulsive food behavior and decision-making based on convenience (Loewenstein, 1996).

Status quo framing occurs when a choice that is typically an alternative is placed in the status quo position while other choices become alternative options (Samuelson and Zackhauser, 1988). The "status quo" choice becomes a more frequent choice than it would have been had it simply been one of the alternatives. Related to this is the "default option," which is the option provided if the consumer does not choose otherwise. Defaults can often become the status quo. At a fast-food restaurant, customers tend to stick to the default options such as fries or chips despite the availability of healthy options such as a salad or fruit (Kahneman, 2003). In addition, customers tend to drink sugary sodas even when other beverages such as water are available and can be self-dispensed. In addition to being unhealthy, these default options often come in large portion sizes. Outside of the fast-food arena, large portion sizes are also problematic in restaurants, supermarkets, and at home (Wansink et al., 2006).

Interventions in the field of behavioral economics should supplement findings from previous economic research. Since traditional economic research focuses primarily on prices, and supply and demand, and taxes on eating decisions, behavioral economics may provide further insight on the psychological, cognitive, and emotional influences on eating behaviors. Although not an alternative to the existing interventions, behavioral economics may provide valuable information for effective policy change to reduce the availability and promotion of foods high in sugar and fat.

Behavioral economics approaches can complement regulation, bans, and taxation

Making healthy dietary choices is difficult in the current food environment, where unhealthy food is typically inexpensive, highly palatable, and widely available. Consumers' decisions specific to food are often driven by what they find most appealing (Lieberman, 2006) as opposed to which foods are best for maintaining good health. Policymakers and public health professionals alike have pursued food policy changes to encourage food and beverage industries to provide more healthful options for consumers. Some of these policies, such as the Sugary Drinks Portion Cap Rule that was proposed in New York City, could be considered as inspired by behavioral economics approaches. They attempt to nudge consumers toward smaller portions but do not restrict the number of portions that can be purchased and consumed. Consumers who are willing to purchase two drinks, for example, could circumvent the policy, though it's possible that many consumers would adjust to the new default portion. Despite the fact that the Sugary Drinks Portion Cap Rule would not force consumers to limit their consumption, the policy was characterized as a "restriction" by the industry, which helped them lobby against the proposal and prevent its passage.

The sharp contrast between the ways industry and public health experts frame the issue is a strong indicator of the political nature of food policy considerations.

The behavioral economics model brings to light the concept of *behavioral nudge interventions*, which can be utilized in the food policy realm as tools to shift behavior toward healthier alternatives (Roberto and Kawachi, 2016). When behavioral nudge principles are used to inform the design and implementation of food policies, they can serve to adjust the environment while preserving food choices. These approaches can either supplement or replace proposals for regulation, taxation, or bans (Rayner and Lang, 2011). However, it is critical that "soft" nudges do not substitute for regulations or primary and secondary prevention efforts that show strong empirical effectiveness for changing behavior. Indeed, these nudges may be most useful in designing how best to implement more intensive regulations. For example, making 90 percent of beverages low- or no-calorie beverage items in workplace vending machines could shift consumers toward healthier items while not completely removing the possibility of purchasing unhealthy items. Nudging interventions promote the healthy behaviors that consumers want for themselves yet still allow them to make choices for themselves (Thaler and Sunstein, 2003).

Moreover, the use of *implementation intentions*, which are defined as predetermined, goal-directed responses to situations that typically provoke unhealthy behaviors (Gollwitzer and Sheeran, 2006), is another alternative to regulation. Although implementation intentions are not entirely effective in changing habits, research has demonstrated that linking certain situations to desired responses can prove effective in working toward behavior change when consumers have specific goals (Gollwitzer and Sheeran, 2006; Orbell and Sheeran, 2000; Sheeran and Orbell, 1999). For example, a group of researchers asked participants to form an implementation intention as an alternative to an unhealthy habit (e.g., eating fruit instead of chips while watching TV) in a situation where they would normally perform an unhealthy behavior. Results showed that responses to implementation intention cues were strengthened while responses to habitual cues were weakened, showing that participants were more capable of thinking of the healthy behaviors (Adriaanse et al., 2011).

Ways for the food industry to use convenience to improve health

Convenience is an important factor for most Americans when making decisions, particularly when choosing foods and beverages (Glanz et al., 1998; Chou et al., 2004; Cutler et al., 2003; Rozin, 2005). Free soda refills are a useful example of how convenience affects behavior. Specifically, the convenience of free refills allows customers to skip the line and have more of the beverage to quench their thirst without having to pay additional money. Similarly, one study demonstrated that when students were required to use a separate checkout lane to purchase potato chips as an accompaniment to their meal, they were less likely to make the purchase (Meiselman et al., 1994). The same idea applies to full-service restaurants and fast-food outlets with drive-thru windows where customers often do not have to wait a long time before receiving their food and can quickly continue on to their next destination. In fact, Chou, Grossman, and Saffer (2004) conclude that convenience based on location of these fast-food and full-service restaurants serves as the primary predictor of local obesity trends. Food companies could harness the appeal of convenience through a number of creative methods, including faster checkout lines for ordering healthy items at fast-food venues, placing drive-through produce windows at grocery stores and pharmacies where food is sold, or showcasing more healthy options at the checkout area in supermarkets to replace the candy items often found in checkout aisles.

Leveraging default options to promote health in restaurants and grocery stores

Research has demonstrated that consumers tend to stick with default options when making menu elections, regardless of how the alternative options are presented to them (Kahneman, 2003). The food and beverage industry can easily partake in encouraging the healthy eating behavior of their patrons by offering healthy options as the default choice for menu items. Thus, the typical unhealthy items that are offered as the default would serve as alternative options that would be available to the consumer upon request. One study found an increase in purchases of healthy foods and a decrease in purchases of unhealthy foods when only healthy items were shown in a cafeteria convenience line (Hanks, 2012). Other studies have also demonstrated that consumers will be more inclined to choose healthier food items when they are more accessible (van Kleef et al., 2012). This ultimately resulted in a decrease in potato chip sales. By manipulating environments to encourage healthy eating behaviors, consumers are more inclined to make healthier food choices.

Larger portion sizes have become increasingly prominent in both restaurant and fast-food meals (Nielsen and Popkin, 2003), as well as in items available for purchase in grocery stores and supermarkets (Wansink, 2006), and portion size has been noted as a contributor of overeating amongst Americans (Rolls, 2003). Such portion sizes can induce consumers to purchase their foods more often while providing more caloric and food "bang" for their buck. They can often also charge more for such larger portions and such that their marginal costs are small (for example, on things like sugary beverages) this can increase profits. Applying the idea of mass defaults to decrease the portion size for all customers would ultimately work toward having consumers eat less at a time. Wansink, Painter, and North (2005) report that the portion size consumers are given may imply what is perceived as normal. Thus, if consumers were provided with smaller portions, they would be more inclined to see this as normal and consume less. This could ultimately reduce overconsumption.

Using food cues to promote healthy dietary choices

Television commercials, billboards, product placement in movies, and food placed strategically in checkout lines are all external cues that prompt individuals to consume food. Consumers' internal cues (e.g., feelings of satiety) can be overwhelmed by these external cues. One study found that participants who rely on internal cues (e.g., fullness) consumed significantly more calories when they were primed with food cues compared to those who did not receive a food cue (Liu et al., 2014). Research has suggested that the visceral nature of food makes it all the more challenging for a person to refrain from eating or purchasing it upon being faced with food cues. Companies like McDonald's have already shifted some of their marketing techniques to promote fruit instead of French fries in children's meals, suggesting healthy cues can feasibly replace at least some of the unhealthy cues in fast-food venues.

Additionally, the food and beverage industry can implement name and product description changes to their items in order to enhance the preference for healthier food items. As noted in previous literature, researchers have seen stronger preferences for healthier products develop upon describing the items with adjectives such as "succulent" or "homemade" to make them more attractive (Wansink et al., 2005). For example, The Healthy Lunchroom Initiative saw an increase of up to 26% in fruit consumption when the fruit was described with more appealing language (Wansink, 2006). Furthering this idea, the industry must be cautious when settling on descriptions and names for products so that they do not encourage overconsumption. One study found that dieters perceived a salad to be extremely healthy based on the language used, thus ate more of it (Irmak et al., 2011). One notable example of industry capitalizing on the healthfulness of meals involves McDonald's inclusion of fruit in its Happy Meals. Though fries are still served with the fruit, McDonald's promotes the message that they have sold billions of pieces of fruit in Happy Meals since 2014. Their promotion of salads also represents a marketing shift, though the majority of ads and menus still promote energy-dense, nutrient-poor foods.

Food policy implications associated with behavioral economics approaches

A number of regulatory and voluntary pledges (e.g., non-enforceable corporate commitments to improve marketing standards or healthfulness of menu items) have been implemented in restaurants, school cafeterias, and workplace settings to encourage healthier behaviors. For a thorough discussion of policy recommendations regarding schools and children as compared to suggestions for restaurants and adults, refer to the work of Liu and colleagues (2014). Their policy recommendations account for the complex nuances that allow schools and child-relevant settings to be amenable to food policy changes because children do not have the same choice sets as adults and are embedded in a system that is intended to place at least some value on children's health, whereas restaurant settings (which serve many adult consumers) are more likely to oppose regulatory enforcements that limit freedom of choice and may be more amenable to voluntary changes or incentive-based programs. Specifically, Liu and colleagues (2014) describe how public schools have more flexibility when implementing policies because they are public institutions that are ideal for public policy experiments; society is more willing to regulate the behavior of children (as compared to adults); and government can maintain a higher degree of control over school food environments. School environments may also be ideal places to implement restrictions on child-targeted food marketing, and society may view restrictions in this arena as more palatable than broad-sweeping restrictions on marketing.

In contrast to school environments, Liu and colleagues (2014) point out that regulating restaurants is considerably more challenging for a variety of reasons. Public schools participate in free and reduced school lunch programs and must be responsive to federal regulations. Adult consumers are much more resistant to policies that they perceive as restrictions on freedom and choices, making restaurants less likely to implement changes seen in some school environments. Liu and colleagues propose that government policies could provide performance-based incentives or penalties for serving high proportions of healthy or unhealthy foods. One option could be to provide various levels of tax reduction incentives that correspond to various healthfulness standards (e.g., certain percentages of food and beverage items meet healthy nutrition standards in both nutrient content and portion size). Liu and colleagues (2014) also propose penalties for selling high proportions of unhealthy foods, similar to gas emission penalties. Other options for promoting healthier consumption at restaurants include heavily promoting healthy foods or inviting consumers to downsize their portion.

Other sectors that may be suitable outlets for behavioral change interventions are venues associated with health-related activities (e.g., hospitals or health insurance agencies) or those that supply some foods and beverages to their large workforce (e.g., large corporate firms). Hospital systems and health insurance companies are inherently tied to health, and hospitals in particular face a critical need to consider the health and nutritional wellness of their patients and employees. Indeed, a healthcare system in Portland, Oregon recently decided to prohibit the sale of soda (Hallman, 2015), and other hospitals have closed fast-food restaurants because of the "inconsistent message" to patients about health (Gordon, 2012). While hospitals may experience unique opportunities to shift their food environments, they may set an ideal example for other venues with looser associations to health (e.g., sports arenas), and help build momentum toward improving food options. Furthermore, some companies already supply employees with access to free beverages or meal vouchers when employees stay late. These types of companies may be especially eager to implement behavioral change interventions (e.g., restructured cafeterias) that will improve or maintain the health of their workforce, thus contributing to their overall productivity.

Conclusion

We reviewed traditional and behavioral economics principles that hold promise as dietary intake improvement efforts, especially in the realm of policy change. Though some policies based on traditional economic principles are linked with decreased consumption of unhealthy products (Colchero et al., 2016; Rayner and Lang, 2011; Bassett et al., 2008; Bollinger et al., 2011; Burton et al., 2006; Chu et al., 2009; Pulos and Leng, 2010; Roberto et al., 2010; Variyam and Cawley, 2006), these approaches often assume consumers behave rationally and act according to their best interests. Calorie labeling on menus, SSB taxes, and even mandatory changes to school lunch programs are some examples of traditional policies that typically face intense opposition from the food industry and many consumers because of the perception that some of these policies promote excessive government intervention and restrict freedom of choice. However, as described by Liu and colleagues (2014), interventions that target children or take place in schools are much more palatable and feasible than policies that restrict choices among adults.

Behavioral economics principles consider that consumers often behave in irrational ways, even when perfect information is presented. Consumers' own biases, visceral factors, preferences for immediate gratification, current norms (e.g., mass defaults and personalized defaults in the environment), and a variety of other factors affect the decisions consumers make about food and beverage consumption and physical activity. As suggested by other researchers, behavioral economics approaches (e.g., offering healthy default options for menu items, behavioral nudge intentions toward healthier alternatives to enhance healthy behaviors, changing the ordering of menus as opposed to just offering menu labels) should be used to supplement current and future obesity intervention policies (Liu et al., 2014; Loewenstein and Ubel, 2011). It is critical that policymakers continue to pursue policies rooted in traditional economic policies (e.g., SSB taxes) because these policies may be more effective at reducing consumption (Colchero et al., 2016; Rayner and Lang, 2012).

Future research should examine ways to combine traditional economic approaches (e.g., price incentives) with behavioral nudges or cues that encourage the development of healthy dietary habits. For example, one study demonstrated that taxing unhealthy foods might be more effective if a label explains that the product is being taxed because it is unhealthy (Shah et al., 2013). One critical area of study involves examining how behavioral economics principles affect low-income consumers or individuals with limited literacy. Finally, existing food assistance programs that already involve significant government oversight (e.g., the Supplemental Nutrition Assistance Program; local government food procurement policies) may be ideal settings to test behavioral economics principles described in this chapter.

Taken together, these concepts from traditional and behavioral economics research can be leveraged to form smarter policies that make it easy and convenient for consumers to engage in healthy behaviors. A number of current obesity prevention and intervention policies are implemented without rigorous empirical testing, which is critical for ensuring the identification of proven, effective programs that can be replicated. Furthermore, the implementation of untested interventions may inadvertently suggest that all environmental approaches to improve dietary intake are ineffective. Behavioral economics principles have the potential to provide helpful supplements to obesity prevention efforts, but policy proposals should be closely tied with empirical research examining the impact of the policy on consumer behavior.

References

Adriaanse, M.A., Gollwitzer, P.M., De Ridder, D.T., De Wit, J.B., and Kroese, F.M. (2011). "Breaking Habits with Implementation Intentions: A Test of Underlying Processes," *Personality and Social Psychology Bulletin* 37(4): 502–513.

Bassett, M.T., Dumanovsky, T., Huang, C., Silver, L.D., Young, C., Nonas, C., Matte, T.D., Chideya, S., and Frieden, T.R. (2008). "Purchasing Behavior and Calorie Information at Fast-Food Chains in New York City, 2007," *American Journal of Public Health* 98(8): 1457–1459.

Bollinger, B., Leslie, P., and Sorensen, A. (2011). "Calorie Posting in Chain Restaurants," *American Economic Journal: Economic Policy* 3(1): 91–128.

Booth, K.M., Pinkston, M.M., and Poston, W.S.C. (2005). "Obesity and the Built Environment," *Journal of the American Dietetic Association* 105(5): 110–117.

Brownell, K.D. and Pomeranz, J.L. (2014). "The Trans-Fat Ban—Food Regulation and Long-Term Health," *New England Journal of Medicine* 370(19): 1773–1775.

Burton, S. and Kees, J. (2012). "Flies in the Ointment? Addressing Potential Impediments to Population-Based Health Benefits of Restaurant Menu Labeling Initiatives," *Journal of Public Policy & Marketing* 31(2): 232–239.

Cantor, J., Torres, A., Abrams, C., and Elbel, B. (2015). "Five Years Later: Awareness of New York City's Calorie Labels Declined, With No Changes in Calories Purchased," *Health Affairs* 34(11): 1893–1900.

Chou, S.Y., Grossman, M., and Saffer, H. (2004). "An Economic Analysis of Adult Obesity: Results from the Behavioral Risk Factor Surveillance System," *Journal of Health Economics* 23(3): 565–587.

Chu, Y.H., Frongillo, E.A., Jones, S.J., and Kaye, G.L. (2009). "Improving Patrons' Meal Selections Through the Use of Point-of-Selection Nutrition Labels," *American Journal of Public Health* 99(11): 2001–2005.

Colchero, M.A., Popkin, B.M., Rivera, J.A., and Ng, S.W. (2016). "Beverage Purchases from Stores in Mexico Under the Excise Tax on Sugar Sweetened Beverages: Observational Study," *British Medical Journal* 352: h6704.

Cutler, D., Glaeser, E., and Shapiro, J. (2003). "Why Have Americans Become More Obese?" *National Bureau of Economic Research*. Available at https://dash.harvard.edu/bitstream/handle/1/2640583/cutler_obese.pdf?sequence=2 (accessed January 4, 2017).

Darmon, N., Ferguson, E.L., and Briend, A. (2002). "A Cost Constraint Alone Has Adverse Effects on Food Selection and Nutrient Density: An Analysis of Human Diets by Linear Programming," *The Journal of Nutrition* 132(12): 3764–3771.

Drewnowski, A. and Specter, S.E. (2004). "Poverty and Obesity: The Role of Energy Density and Energy Costs," *The American Journal of Clinical Nutrition* 79(1): 6–16.

Frederick, S., Loewenstein, G., and O'Donoghue, T. (2002). "Time Discounting and Time Preference: A Critical Review," *Journal of Economic Literature* 40(2): 351–401.

French, S.A. and Story, M. (2013). "Commentary on Nutrition Standards in the National School Lunch and Breakfast programs," *JAMA Pediatrics* 167(1): 8–9.

Glanz, K., Basil, M., Maibach, E., Goldberg, J., and Snyder, D.A.N. (1998). "Why Do Americans Eat What They Do?: Taste, Nutrition, Cost, Convenience, and Weight Control Concerns as Influences on Consumption," *Journal of the American Dietetic Association* 98(10): 1118–1126.

Goldstein, D.G., Johnson, E.J., Herrmann, A., and Heitman, M. (2008) "Nudge Your Customers Toward Better Choices," *Harvard Business Review* 1 Dec. Web. 30 Oct. 2015.

Gollwitzer, P.M. and Sheeran, P. (2006). "Implementation Intentions and Goal Achievement: A Meta-analysis of Effects and Processes. *Advances in Experimental Social Psychology* 38: 69–119.

Gordon, K. (2012). "Fast Food Chains in Cafeterias Put Hospitals in a Bind," Web Site. Available at http://www.npr.org/sections/thesalt/2012/04/05/150091951/fast-food-chains-in-cafeterias-put-hospitals-in-a-bind (accessed January 4, 2017).

Hallman, T. (2015). "Oregon Hospital System Providence Bans the Sale of Sugary Drinks," Web Site. Available at http://www.oregonlive.com/portland/index.ssf/2015/01/portland_hospital_system_bans.html. (accessed January 4, 2017).

Hanks, A.S., Just, D.R., Smith, L.E., and Wansink, B. (2012). "Healthy Convenience: Nudging Students Toward Healthier Choices in the Lunchroom," *Journal of Public Health* 34(3): 370–376.

Hoyt, L.T., Kushi, L.H., Leung, C.W., Nickleach, D.C., Adler, N., Laraia, B.A., Hiatt, R.A., and Yen, I.H. (2014). "Neighborhood Influences on Girls' Obesity Risk Across the Transition to Adolescence," *Pediatrics* 134(5): 942–949.

Hu, F.B. and Willett, W.C. (2002). "Optimal Diets for Prevention of Coronary Heart Disease," *The Journal of the American Medical Association* 288(20): 2569–2578.

Irmak, C., Vallen, B., and Robinson, S.R. (2011). "The Impact of Product Name on Dieters' and Nondieters' Food Evaluations and Consumption," *Journal of Consumer Research* 38(2): 390–405.

Kahneman, D. (2003). "Maps of Bounded Rationality: Psychology for Behavioral Economics," *American Economic Review,* 1449–1475.

Kruger, D.J., Greenberg, E., Murphy, J.B., DiFazio, L.A., and Youra, K.R. (2014). "Local Concentration of Fast-Food Outlets Is Associated with Poor Nutrition and Obesity," *American Journal of Health Promotion* 28(5): 340–343.

Leonard, T., Shuval, K., de Oliveira, A., Skinner, C.S., Eckel, C., and Murdoch, J.C. (2013). "Health Behavior and Behavioral Economics: Economic Preferences and Physical Activity Stages of Change in a Low-Income African-American Community," *American Journal of Health Promotion* 27(4): 211–221. doi: 10.4278/ajhp.110624-QUAN-264.

Liu, P.J., Wisdom, J., Roberto, C.A., Liu, L.J., and Ubel, P.A. (2014). "Using Behavioral Economics to Design More Effective Food Policies to Address Obesity," *Applied Economic Perspectives and Management* 36: 6–24.

Loewenstein, G. (1996). "Out of Control: Visceral Influences on Behavior," *Organizational Behavior and Human Decision Processes* 65(3): 272–292. Web.

Loewenstein G. and Ubel, P. (2010). "Economics Behaving Badly," *The New York Times.* Web Site. Available at http://bear.warrington.ufl.edu/williams/MAR_6930/Readings_files/Loewenstein%20%26%20Ubel.pdf (accessed January 4, 2017).

Malik, V.S., Willett, W.C., and Hu, F.B. (2013). "Global Obesity: Trends, Risk Factors and Policy Implications," *Nature Reviews Endocrinology* 9(1): 13–27.

Mandaro, L. (2014). "Nation's First Soda Tax Is Passed," Web Site. Available at http://www.usatoday.com/story/news/nation-now/2014/11/05/berkeley-passes-soda-tax/18521923/ (accessed January 4, 2017).

Mayne, S.L., Auchincloss, A.H., and Michael, Y.L. (2015). "Impact of Policy and Built Environment Changes on Obesity-Related Outcomes: A Systematic Review of Naturally Occurring Experiments," *Obesity Reviews* 16(5): 362–375.

Meiselman, H.L., Hedderley, D., Staddon, S.L., Pierson, B.J., and Symonds, C.R. (1994). "Effect of Effort on Meal Selection and Meal Acceptability in a Student Cafeteria," *Appetite* 23(1): 43–55. Available at http://www.ncbi.nlm.nih.gov/pubmed/7826056 (accessed January 4, 2017).

Miller, W.C, Koceja, D.M., and Hamilton E.J. (1997). "A Meta-Analysis of the Past 25 Years of Weight Loss Research Using Diet, Exercise or Diet Plus Exercise Intervention," *International Journal of Obesity* 21(10): 941–947.

Nielsen, S.J. and Popkin, B.M. (2003). "Patterns and Trends in Food Portion Sizes," *Journal of the American Medical Association* 289(4): 450–453.

O'Donoghue, T. and Rabin, M. (2000). "The Economics of Immediate Gratification," *Journal of Behavioral Decision Making* 13(2): 233–250.

Ogden, C.L., Flegal, K.M., Carroll, M.D., and Johnson, C.L. (2002). "Prevalence and Trends in Overweight Among US Children and Adolescents, 1999–2000," *JAMA* 288(14): 1728–1732.

Ogden, C.L., Carroll, M.D., Kit, B.K., and Flegal, K.M. (2012). "Prevalence of Obesity and Trends in Body Mass Index Among US Children and Adolescents, 1999–2010," *JAMA* 307(5): 483–490.

Ogden, C.L., Carroll, M.D., Kit, B.K., and Flegal, K.M. (2014). "Prevalence of Childhood and Adult Obesity in the United States, 2011–2012," *JAMA* 311(8): 806–814.

Orbell, S. and Sheeran, P. (2000). "Motivational and Volitional Processes in Action Initiation: A Field Study of the Role of Implementation Intentions," *Journal of Applied Social Psychology* 30(4): 780–797. doi:19111/j.1559-1816.2000.tb02823.x.

Parker, L., Burns, A.C., and Sanchez, E. (eds) (2009). *Local Government Actions to Prevent Childhood Obesity.* Washington, DC: National Academies Press.

Pulos, E. and Leng, K. (2010). "Evaluation of a Voluntary Menu-Labeling Program in Full-Service Restaurants," *American Journal of Public Health* 100(6): 1035–1039.

Putnam, G.P. (1838). *The Tourist in Europe.* New York: Wiley & Putnam.

Rayner, G. and Lang, T. (2011). "Is Nudge an Effective Public Health Strategy to Tackle Obesity? No," *British Medical Journal* 342: d2177.

Rayner, G. and Lang, T. (2012). "Ecological Public Health," *Health of People, Places and Planet*, 617.

Roberto, C.A. and Kawachi, I. (2016). *Behavioral Economics & Public Health.* New York, NY: Oxford University Press.

Roberto, C.A., Larsen, P.D., Agnew, H., Baik, J., and Brownell, K.D. (2010). "Evaluating the Impact of Menu Labeling on Food Choices and Intake," *American Journal of Public Health* 100(2): 312–318.

Rolls, B.J. (2003). "The Supersizing of America: Portion Size and the Obesity Epidemic," *Nutrition Today* 38(2): 42–53.

Rose, M. (2000). "Society: The Emergence of Urban Britain", in C. Haigh (ed.), *The Cambridge Historical Encyclopaedia of Great Britain and Ireland.* Cambridge: Cambridge University Press [1985], pp. 276–281.

Rozin, P. (2005). "The Meaning of Food in Our Lives: A Cross-Cultural Perspective on Eating and Well-Being," *Journal of Nutrition Education and Behavior,* 37: S107–S112.

Sallis, J.F., Floyd, M.F., Rodríguez, D.A., and Saelens, B.E. (2012). "Role of Built Environments in Physical Activity, Obesity, and Cardiovascular Disease," *Circulation* Feb 7, 125(5): 729–737.

Samuelson, W. and Zeckhauser, R. (1988). "Status Quo Bias in Decision Making," *Journal of Risk and Uncertainty* 1(1): 7–59. Web.

Shah, A.M., Bettman, J.R., Keller, P.A., Edell Britton, J.A., and Ubel, P.A. (2013). "The Effect of Sin Taxes and Evaluative Labels on Demand for Unhealthy Products," Paper presented at the Theory+Practice in Marketing Conference, London Business School, London, UK.

Sheeran, P. and Orbell, S. (1999). "Implementation Intentions and Repeated Behaviour: Augmenting the Predictive Validity of the Theory of Planned Behaviour," *European Journal of Neuroscience* 29: 349–369.

Stigler, G.J. (1961). "The Economics of Information," *The Journal of Political Economy* 213–225.

Tavernise, S. (2015). "F.D.A Extends Deadline for Calorie Counts on Menus," *New York Times*. Web Site. Available at http://www.nytimes.com/2015/07/10/health/fda-delays-deadline-for-calorie-count-on-menus.html?rref=collection%2Fbyline%2Fsabrina-tavernise&action=click&contentCollection=undefined&region=stream&module=stream_unit&version=search&contentPlacement=3&pgtype=collection (accessed January 4, 2017).

Thaler, R.H. and Sunstein, C.R. (2003). "Libertarian Paternalism," *American Economic Review Papers and Proceedings* 93: 175–179.

Tsai, A.G. and Wadden, T.A. (2005). "Systematic Review: An Evaluation of Major Commercial Weight Loss Programs in the United States," *Annals of Internal Medicine* 142(1): 56–66.

van Kleef, E., Otten, K., and van Trijp, H.C. (2012). "Healthy Snacks at the Checkout Counter: A Lab and Field Study on the Impact of Shelf Arrangement and Assortment Structure on Consumer Choices," *BMC Public Health* 12(1): 1072.

Variyam, J.N. and Cawley, J. (2006). *Nutrition Labels and Obesity* (No. w11956). National Bureau of Economic Research.

Wadden, T.A., Butryn, M.L., and Byrne, K.J. (2004). "Efficacy of Lifestyle Modification for Long⊠Term Weight Control," *Obesity Research* 12(S12): 151S–162S.

Wansink, B. (2006). "Mindless Eating: Why We Eat More Than We Think," New York, NY: Bantam Dell.

Wansink, B., Painter, J.E., and North, J. (2005). "Bottomless Bowls: Why Visual Cues of Portion Size May Influence Intake," *Obesity Research* 13(1): 93–100.

Wansink, B., Van Ittersum, K., and Painter, J.E. (2005). "How Descriptive Food Names Bias Sensory Perceptions in Restaurants," *Food Quality and Preference* 16(5): 393–400.

Wansink, B., Van Ittersum, K., and Painter, J.E. (2006). "Ice Cream Illusions," *American Journal of Preventive Medicine* 31(3): 240–243. Web.

Weiss, S., Davis, E., Wojtanowski, A.C., Foster, G.D., Glanz, K., and Karpyn, A. (2015). "Consumer Taste Tests and Milk Preference in Low-Income, Urban Supermarkets," *Public Health Nutrition* 18(08): 1419–1422.

Willett, W.C. and Skerrett, P.J. (2001). *Eat, Drink, and Be Healthy: The Harvard Medical School Guide to Healthy Eating.* New York, NY: Simon & Schuster, Inc. pp. 22–23.

Winterer, C. (2002). *The Culture of Classicism: Ancient Greece and Rome in American Intellectual Life, 1780–1910.* Baltimore: Johns Hopkins University Press.

Woo Baidal, J.A. and Taveras, E.M. (2014). "Protecting Progress Against Childhood Obesity—the National School Lunch Program," *New England Journal of Medicine* 371(20): 1862–1865.

Young, L.R. and Nestle, M. (2002). "The Contribution of Expanding Portion Sizes to the US Obesity Epidemic," *American Journal of Public Health* 92(2): 246–249. Web.

Young, M.D., Morgan, P.J., Plotnikoff, R.C., Callister, R., and Collins, C.E. (2012). "Effectiveness of Male-Only Weight Loss and Weight Loss Maintenance Interventions: A Systematic Review with Meta-analysis," *Obesity Reviews* 13(5): 393–408.

PART III

Detecting and managing disease

7

IMPROVING MEDICATION ADHERENCE WITH BEHAVIORAL ECONOMICS

Steven E. Meredith and Nancy M. Petry

> *Increasing the effectiveness of adherence interventions may have a far greater impact on the health of the population than any improvement in specific medical treatments.*
>
> —R. Brian Haynes (2001)

Introduction

Nearly half of all Americans take at least one prescription medication (Centers for Disease Control and Prevention, 2015), and this proportion is expected to increase as the US population ages. Although medications have been used to improve the health and wellbeing of millions of individuals around the world, drugs only work for patients who take them. Nonadherence leads to poor health outcomes, increased mortality, disease spread, and avoidable health care costs, including $100 billion annually in excess hospitalizations (Osterberg and Blaschke, 2005).

Some estimates suggest as many as 28 percent of new prescriptions in the US are never filled (Fischer et al., 2010), and for patients who fill their prescriptions, nonadherence may still become a problem even after they leave the pharmacy. Patients must then take the correct number of doses at the appropriate interval and refill their prescriptions as often and for as long as their health condition requires. Thus, for any given patient, the rate of medication adherence can vary widely from 0 percent to over 100 percent (if the patient takes more than the amount prescribed).

Although taking too much medication is problematic, patients typically do not take enough. Rates of medication adherence must be relatively high for patients to reap the full therapeutic benefits of prescription drugs. For instance, in a study that was conducted prior to recent advances in antiretroviral therapy, Paterson et al. (2000) found patients who were prescribed protease inhibitors for

HIV needed to achieve adherence rates of 95 percent or higher to suppress viral replication. Regardless of the disease or drug, however, patients are generally considered adherent if they take at least 80 percent of their medication, and for many drugs, this level of adherence can produce significant health benefits (e.g., Wei et al., 2002).

Although most patients are unlikely to achieve optimal levels of medication adherence, estimating the average rate of adherence among all patients is challenging because rates vary by medication and participant characteristics. Further, measures of adherence are inconsistent across studies (e.g., self-report, pill count, or pharmacy claims data). Nevertheless, most research suggests many patients stop taking medications within a few months after filling their first prescription, and average rates of adherence typically range from 50 to 75 percent (Cramer and Rosenheck, 1998; DiMatteo, 2004; Osterberg and Blaschke, 2005).

Medication adherence and chronic disease

Medication adherence is critically important to treating all types of medical problems, including acute health conditions such as those requiring antibiotics. However, adherence is often lowest among patients with chronic diseases. Recent estimates indicate about half of US adults suffer from at least one chronic disease (Ward et al., 2014), and a 2003 report by the World Health Organization suggests adherence to therapies for chronic diseases is roughly 50 percent in developed countries and even lower in developing countries (Sabaté, 2003).

For patients with chronic health conditions, research shows poor medication adherence leads to increased service utilization, diagnostic testing, treatment for comorbid conditions, hospitalization, and, consequently, higher health care costs. For example, in a population-based sample of over 137,000 patients, Sokol and colleagues (2005) found significantly higher hospitalization rates in nonadherent patients with diabetes, hypertension, congestive heart failure, and high cholesterol relative to adherent patients with the same chronic health conditions. In addition, they found disease-related health care costs were significantly higher among nonadherent patients who had diabetes and high cholesterol relative to adherent patients with the same conditions.

To make matters worse, poor adherence among patients with chronic diseases is also associated with increased mortality (Simpson et al., 2006). For example, in a study of patients with heart disease and diabetes, Ho and colleagues found patients who were nonadherent to their cardioprotective medication (e.g., angiotensin-converting enzyme inhibitors and statins) had a significantly higher rate of mortality than adherent patients (12.1% vs. 6.7%; Ho et al., 2006).

Barriers to medication adherence

The problem of nonadherence is widespread and has been attributed to a variety of factors (Brown and Bussell, 2011). According to a survey of nearly 10,000 US

TABLE 7.1 Barriers to medication adherence

Patient Level	Financial	Low income No health insurance Lack of transportation to medical appointments or pharmacy
	Mental/cognitive	Forgetting to take medication Poor health literacy Cognitive impairment Substance abuse Serious psychiatric illness
	Negative beliefs/ attitudes	Lack of faith in treatment effectiveness Belief that an asymptomatic disease does not require treatment Misinformation about medication or disease
Medication/ Provider Level	Financial	High drug costs High copayments
	Other	Negative side-effects of medication Multiple or complex medication regimens Lack of follow-up Lack of patient education

patients, the most commonly reported reasons for missing a dose of medication are forgetting to take the medication, avoiding (real or perceived) side effects, the high cost of the medication, and believing the medication will have little or no effect on health (Boston Consulting Group, 2003). Table 7.1 lists some of the most common barriers to medication adherence.

Importantly, many barriers to adherence are surmountable, and researchers, practitioners, and policymakers have developed and tested various interventions designed to help patients overcome these barriers. The remainder of this chapter will review some of these medication adherence interventions, with a special focus on reinforcement-based behavioral economics interventions.

Increasing medication adherence

Although some studies have shown patient education (Lowe et al., 2000), case management (Piette et al., 2000), and pharmaceutical counseling (Bouvy et al., 2003) can help increase medication adherence, authors of a recent Cochrane review of medication adherence interventions concluded, "Current methods of improving medication adherence for chronic health problems are mostly complex and not very effective" (p. 2; Nieuwlaat et al., 2014). Many of these interventions involve family or peer support combined with comprehensive education and counseling from health care professionals. Because these complex interventions often produce only modest improvements in adherence, some researchers have investigated a simpler approach to improving medication adherence: minimizing or eliminating financial barriers.

Increasing adherence with traditional economics

The high cost of prescription drugs and insurance copayments are major barriers to adherence. In a nationwide survey of Medicare beneficiaries conducted in 2004, the most frequently cited reason for not filling a prescription was, "thought it would cost too much" (56%), followed by, "medicine not covered by insurance" (20%; Kennedy et al., 2008). Further, several studies have reported inverse relations between medication adherence and the cost of copayments.

Huskamp and colleagues (2003) examined prescription drug use among participants enrolled in an employer-sponsored health plan before and after the employer increased copayments for certain brand-name drugs. After the change was implemented, enrollees were significantly more likely to switch to less expensive generic or brand-name drugs than employees in a control group. Moreover, many enrollees simply stopped refilling their prescriptions rather than switching to more affordable medications. Thus, increases in copayments were associated with decreases in medication adherence.

Research has also shown decreases in copayments are associated with increases in adherence. After Blue Cross Blue Shield of North Carolina reduced copayments for brand-name drugs and eliminated copayments for generic drugs, medication adherence significantly improved for patients with diabetes, high cholesterol, hypertension, and congestive heart failure (Maciejewski et al., 2010).

This relationship between cost sharing and medication adherence is remarkably systematic. A comprehensive review of 132 articles examining associations between medication adherence and various cost sharing features of prescription drug benefits showed increased cost sharing was associated with nonadherence, and for every 10 percent increase in patient cost sharing, prescription drug spending decreased by 2 percent to 6 percent (Goldman et al., 2007).

Taken together, data from these studies show the demand for prescription medications is price sensitive. Thus, armed with traditional economic theory, an increasing number of employers and insurers are combatting medication nonadherence with value-based insurance designs (VBIDs) that minimize or eliminate financial barriers. These cost-sharing plans require lower copayments for medications that provide greater therapeutic benefits (Chernew et al., 2007).

Thus far, research suggests VBIDs are associated with statistically significant increases in medication adherence among patients with chronic health conditions (e.g., Farley et al., 2012). Notably, however, most studies report only modest (1% to 5%) increases in adherence to targeted drugs. Further, the long-term effects of VBIDs on morbidity, mortality, and health care costs remain unclear. Nevertheless, if this traditional economic approach to promoting medication adherence is adopted nationwide, then small improvements in adherence could have major public health repercussions.

Although medication adherence interventions based on traditional economic theory have produced promising, albeit modest, improvements in adherence, there is certainly room for additional improvement. Consequently, an increasing number of medication adherence interventions are emerging from the field of behavioral economics.

Increasing adherence with behavioral economics

Traditional economics (or rational choice) theory predicts patients will adhere to their prescriptions when the long-term benefits of doing so outweigh the short-term costs. In other words, this theory suggests patients will pay a relatively small monetary fee today for a medication that will produce relatively large health benefits in the future. However, this theoretical perspective does not take into account the behavioral economics of intertemporal choice (i.e., the influence of time on decision-making) or, more specifically, delay discounting (i.e., the tendency to discount the value of future consequences relative to present consequences).

Medication adherence is operant behavior (i.e., behavior maintained and modifiable by its consequences), and like other health-related operant behavior, delay discounting plays an important role in adherence. A brief appraisal of the immediate and delayed consequences associated with medication adherence reveals why this behavior is an ideal target for behavioral economics intervention.

The short-term beneficial consequences of medication adherence often go unnoticed (e.g., if the health condition is asymptomatic or the drug effects are imperceptible). Consequently, patients may not recognize the benefits of adherence (e.g., improved health or longer life) for many weeks, months, or years after they start taking a drug. Due to this delay, patients are likely to discount the benefits of medication adherence, thereby, limiting the influence of these consequences on their behavior. Conversely, the costs associated with adherence, although comparatively modest (e.g., the cost of a copayment, visiting a pharmacy to pick up refills, drug side effects), are more immediate and, therefore, have greater influence on behavior. For some patients, these short-term response costs outweigh the long-term benefits associated with medication adherence, resulting in nonadherence.

Some behavioral economics interventions are designed to shift the balance between the short-term and long-term consequences of adherence by minimizing the response costs associated with the behavior. For example, CVS/caremark™ offers an automated refill program called ReadyFill at Mail® in which CVS pharmacies will refill and mail eligible prescriptions to enrollees prior to the prescription refill date. This type of intervention could help improve medication adherence by minimizing the effort required by patients to refill their prescriptions.

To date, there are limited data on the efficacy and acceptability of ReadyFill at Mail® (Keller et al., 2011). Presumably, however, behavioral economics interventions that minimize or eliminate the response costs associated with refilling a prescription are likely to increase refills. Similarly, traditional economic interventions that minimize or eliminate the financial costs associated with refilling a prescription are also likely to increase refills. Yet, medication adherence involves more than just refilling prescriptions once every one to three months. Patients must also take their medications—often daily. An intervention that targets each of these responses may be necessary to increase adherence for some patients.

Reinforcement-based interventions are behavioral economics interventions that provide proximate reinforcement (e.g., small cash incentives) when patients take a dose of medication as prescribed. Numerous studies have shown reinforcement-based interventions can increase medication adherence in patients suffering from a variety of health conditions. Several examples of these interventions are discussed below and outlined in Table 7.2.

Methadone-based reinforcement of disulfiram adherence

The rate of alcohol abuse is high in opioid-dependent patients, and alcohol use can interfere with methadone maintenance treatment (Ottomanelli, 1999). Disulfiram administration may help reduce alcohol consumption in this population. Disulfiram inhibits acetaldehyde metabolism, causing patients to become ill (i.e., experience symptoms such as headache and nausea) when they consume alcohol. The drug is effective in reducing alcohol consumption when patients are adherent; however, its clinical utility is limited by very low rates of adherence (20% to 50%; Wright and Moore, 1990). Thus, Liebson et al. (1978) developed one of the first reinforcement-based medication adherence interventions in an effort to promote disulfiram adherence (Table 7.2).

In this study, male methadone patients who were recently or soon to be discharged from treatment due to alcohol-related problems were randomly assigned to one of two groups. In a reinforcement group, patients received methadone doses contingent on supervised ingestion of disulfiram. In a control group, patients received methadone regardless of disulfiram adherence. The percentage of patients' drinking days were assessed with breath alcohol tests and found to be much lower in the reinforcement group (2%) relative to the control group (21%). An increased rate of employment and fewer arrests were also observed among patients in the methadone-based reinforcement group. Although disulfiram adherence was not objectively measured in this study, the authors reported that control patients admitted they were not taking their disulfiram.

Results from Liebson et al. (1978) suggested methadone-based reinforcement could be used to promote medication adherence, even to a drug with notoriously low rates of adherence. Subsequent investigations showed methadone-based reinforcement could also be used to improve medication adherence among opioid-dependent patients with other health problems (e.g., Elk et al., 1993). Of course, it may be unethical to deny methadone to patients who refuse to adhere to disulfiram or other medications. Thus, Liebson et al. only enrolled patients who were already at risk for discharge from a methadone treatment program.

TABLE 7.2 Reinforcement-based interventions to promote medication adherence

Reference	*Population*	*Medication*	*Regimen*	*Intervention Duration*	*Reinforcer*	*Adherence Outcome (Reinforcement vs Control)*	*Major Limitations*
Liebson et al. (1978)	Methadone patients	Disulfiram for alcohol abuse	Daily (orally)	6 months	Methadone	—	No objective measure of adherence
Dunn et al. (2013)	People who inject drugs	Naltrexone for opioid dependence	3 days/week (orally)	26 weeks	Employment: up to $10 per hour	72% vs 21%	—
Staring et al. (2009)	Outpatient psychiatric patients	Antipsychotic for schizophrenia	One injection every 2–4 weeks	12 months	Cash: €10–20 per dose	100% vs 44%	Pilot study not designed to evaluate efficacy
Cass et al. (2005)	Patients 14 years of age and younger	Isoniazid for latent tuberculosis	Daily (orally)	9 months	Tokens: 1 sticker per dose; ~30 stickers exchangeable for up to 1 toy each month	92% vs 82%	Inadequate control group and indirect measure of adherence
Rosen et al. (2007)	HIV patients with a history of illicit substance use	Antiretroviral therapy for HIV	At least once daily (orally)	16 weeks	Prizes: 1–21 prize draws per dose exchangeable for up to $800 total in goods or gift cards	76% vs 44%	—
Petry et al. (2015)	Patients with high blood pressure	Antihypertensive medication	Once to twice daily (orally)	12 weeks	Vouchers: $1–6 per dose exchangeable for up to $468 total in goods or gift cards	99% vs 92%	—

Importantly, take-home methadone doses can also be used as reinforcers in methadone clinics (e.g., Stitzer et al., 1977). Patients who are awarded this clinic privilege are given an extra dose of methadone to take home with them so they will not have to return to the clinic to pick up another dose the following day.

Employment-based reinforcement of naltrexone adherence

Naltrexone is an opioid antagonist that blocks the reinforcing effects of heroin and other opioids. Due to the drug's mechanism of action and its lack of abuse potential, it holds considerable promise as a relapse prevention tool for opioid dependent patients. However, evidence for its clinical utility is mixed. Authors of a Cochrane review of oral naltrexone studies noted adherence was too low in most studies to draw any definitive conclusions about the efficacy of the drug (Minozzi et al., 2011). Thus, to determine the clinical utility of naltrexone, methods are needed to improve adherence.

Although some studies have shown monetary-based reinforcement of naltrexone ingestion can promote adherence (e.g., Preston et al., 1999), most interventions were of relatively short duration (e.g., 12 weeks). Given the chronic nature of opioid dependence, interventions of longer duration are needed. However, with increasing duration of reinforcement-based interventions comes increasing cost.

Silverman and colleagues developed an innovative strategy for increasing treatment duration without increasing cost: employment-based reinforcement (Silverman et al., 2001). With employment-based reinforcement, patients are permitted access to a "therapeutic workplace" contingent on objective evidence of target behavior change. Patients who are given access to the workplace typically earn $10 per hour for data entry jobs. This treatment method has been applied to a number of health-related behaviors, including oral naltrexone adherence in unemployed injection drug users (Dunn et al., 2013; Table 7.2).

In Dunn et al. (2013), opioid-dependent patients were inducted onto oral naltrexone after detox and then randomized to one of two 26-week conditions. Patients assigned to an employment-based reinforcement group were given access to the therapeutic workplace contingent on supervised ingestion of oral naltrexone. Patients assigned to a control group received a take-home supply of the drug and were permitted access to the workplace regardless of naltrexone adherence. Urine samples were collected monthly to assess adherence in both groups. The mean percentage of naltrexone-positive urine samples was significantly higher among patients in the reinforcement group (72%) relative to the control group (21%). Moreover, all urine samples were naltrexone-positive for 43 percent of reinforcement patients compared to only 3 percent of control patients.

Although oral naltrexone requires frequent dosing (e.g., three times per week), extended-release injectable formulations of the drug have been developed that require less frequent dosing. For example, Vivotrol® requires injections once every four weeks. A study that examined employment-based reinforcement of adherence to depot injections of Vivotrol® showed the intervention significantly improved adherence relative to controls (74% vs. 26%; DeFulio et al., 2012).

Notably, neither oral nor extended release naltrexone adherence was associated with significantly fewer opioid-positive urine samples in these studies. Thus, more research is needed to determine the clinical utility of naltrexone in preventing relapse in opioid dependent patients, especially polydrug using patients. Nevertheless, these studies show that employment-based reinforcement can substantially improve medication adherence, as rates of naltrexone adherence were three times higher in the employment-based reinforcement groups relative to the control groups.

Cash-based reinforcement of antipsychotic medication adherence

Psychiatric illness is a major barrier to medication adherence (Table 7.1). In a systematic review of adherence studies, the mean rate of medication adherence was 76 percent among patients with physical disorders compared to 58 percent among patients with psychiatric disorders (Cramer and Rosenheck, 1998).

Staring and colleagues conducted a pilot study to evaluate cash-based reinforcement of medication adherence in patients diagnosed with schizophrenia who had a recent history of nonadherence and hospitalization (Staring et al., 2010; Table 7.2). The study was conducted in the Netherlands, and patients received 10 to 20 Euro (approximately 15 to 30 USD at the time the study was conducted) contingent on each depot injection of antipsychotic medication. The magnitude of the reinforcer depended on the frequency of the dosing schedule: €10 when injections occurred every two weeks, €15 when they occurred every three weeks, and €20 when they occurred every four weeks.

Patients in this study accepted 44 percent of injections during the year prior to intervention implementation compared to 100 percent of injections when cash-based reinforcers were available during the second year. Further, mean time spent in the hospital fell from 100 days during year one to just three days during year two. Although this study was not designed to evaluate the efficacy of the intervention, it demonstrated that cash-based reinforcement is feasible for increasing medication adherence in a difficult-to-treat population, and the intervention was associated with remarkably robust improvements in outcome measures.

Several randomized controlled trials have also shown that modest monetary reinforcers can be used to promote adherence to various forms of pharmacotherapy (e.g., Seal et al., 2003; Tulsky et al., 2000). One advantage of monetary reinforcement is it relies on a well-established system of generalized conditioned reinforcement. Where some programmed consequences such as methadone and employment only function as reinforcers for individuals who are experiencing a relevant motivating operation (e.g., opioid withdrawal or unemployment), cash functions as a reinforcer for virtually everyone, regardless of an individual's current state of motivation.

Token-based reinforcement of isoniazid adherence in children

Token economies are among the most common reinforcement-based interventions. They were first developed for use in inpatient psychiatric clinics but have subsequently been adopted to improve a variety of behaviors in a range of settings often among children (Kazdin, 1982). In a token economy, conditioned reinforcers (e.g., coins or stickers) are awarded contingent on desirable behavior change, and after a specified number of tokens are earned, they are exchanged for primary or "back-up" reinforcers (e.g., goods or privileges).

Cass and colleagues evaluated a token-based reinforcement intervention to promote medication adherence in children and adolescents diagnosed with latent tuberculosis (Cass et al., 2005; Table 7.2). Patients were 14 years of age or younger and prescribed a 6- to 9-month regimen of daily isoniazid (a first-line treatment for active and latent tuberculosis). At intake, these patients received a 1-month calendar and 30 stickers. They were instructed to place a sticker on the calendar each day they took their medication. When a completed calendar was returned at monthly clinic visits, patients could select a toy or stuffed animal from a "treasure chest." At each visit, patients received another calendar and a new supply of stickers. Relative to a historical comparison group, patients who received token-based reinforcement were significantly more likely to be considered adherent by attending at least six of their monthly clinic appointments (92% vs. 82%).

Daily oral medication adherence was not objectively measured by Cass et al. (2005). Rather, staff relied on patient self-report. Ideally, reinforcement-based interventions should rely on objective measures of adherence such as supervised medication ingestion. In studies where more objective measures were obtained, methadone-based (Elk et al., 1993) and cash-based reinforcement have been shown to increase isoniazid adherence (Tulsky et al., 2000). Thus, token-based reinforcement should also improve adherence when tokens are awarded contingent upon supervised ingestion of isoniazid or other drugs. If so, then the token economy could be a very promising tool that health care professionals could use in inpatient treatment settings and parents could use at home with nonadherent children.

Prize-based reinforcement of antiretroviral medication adherence

Although objective assessment of adherence is an important feature of reinforcement-based interventions, direct supervision of medication ingestion may be difficult or impossible for outpatient treatment providers. Some medication regimens require multiple doses per day, but patients may be unwilling or unable to travel to a clinic to demonstrate adherence on a daily basis. Thus, to obtain reliable and valid estimates of adherence, some researchers have used medication event monitoring system (MEMS) caps. These caps rely on electronic sensors to record the time and date each time a prescription drug bottle is opened.

Rosen et al. (2007) used MEMS caps to assess adherence to antiretroviral medication in HIV positive patients (Table 7.2). After baseline rates of adherence were assessed, patients who were less than 80 percent adherent to their medication regimens were randomized to one of two 16-week conditions. Patients assigned to a counseling group received weekly individual counseling sessions. Patients assigned to a reinforcement group received weekly counseling plus prize-based contingency management. Prize-based contingency management is an intervention that relies on a probabilistic or intermittent schedule of reinforcement (Petry, 2000). Thus, it may be more affordable than some alternative reinforcement-based interventions, and it is becoming increasingly popular in clinical settings (e.g., Petry et al., 2014).

At counseling sessions, patients in the reinforcement group reviewed MEMS data from the previous week. For each day that all medication doses were taken within an agreed-upon 3-hour window, these patients earned one opportunity to win a prize (i.e., they could draw one slip of paper from a prize bowl), and they earned an additional five prize draws when all doses during the week were taken on time. The majority of the slips of paper in the prize bowl read, "good job" but were not associated with any tangible reinforcement. Over one-fourth of the slips could be exchanged for $1 prizes (e.g., bus tokens and fast-food gift certificates), a smaller proportion could be exchanged for $20 prizes (e.g., clothing and retail gift cards), and less than 1 percent could be exchanged for $100 prizes (e.g., small appliances and electronics).

Relative to baseline measures, mean rates of adherence increased significantly more for patients in the reinforcement group (from 61% to 76%) compared to patients in the counseling group, whose mean rates of adherence actually decreased from 59 percent to 44 percent. Moreover, patients in the reinforcement group were significantly more likely to achieve 95 percent adherence over the course of the 16-week intervention, and they had significantly improved viral loads. Thus, prize-based reinforcement of MEMS-confirmed adherence not only improved medication adherence, it also improved health indices.

Notably, other studies have also demonstrated the feasibility and efficacy of reinforcing MEMS-confirmed medication adherence (e.g., Sorensen et al., 2007). Although this approach is promising, two important limitations must be considered. First, patients can open MEMS caps without ingesting the medication. Thus, they could receive programmed consequences without actually taking their medication. Second, when used in outpatient settings, this strategy may require substantial delays between the target behavior and reinforcer delivery. Behavioral economics research shows that such delays can compromise treatment efficacy (e.g., Lussier et al., 2006). Recently, however, researchers have been developing reinforcement-based interventions that can overcome these limitations by using mobile phone technology to remotely monitor and reinforce medication adherence in real time.

Remote voucher-based reinforcement of antihypertensive medication adherence

Petry et al. (2015) developed and tested a mobile phone-based intervention to promote medication adherence (Table 7.2). Patients with high blood pressure (≥120 mmHg systolic and 80 mmHg diastolic) were invited to participate if they took antihypertensive medication for at least three months and reported missing doses within the past month. Patients were randomized to one of two 12-week conditions. Those assigned to a standard care condition attended a 30-minute session on improving medication adherence and were instructed to visit their physician as usual. Patients assigned to a reinforcement condition received standard care and were instructed to video-record themselves ingesting each dose of antihypertensive medication with a mobile phone and send the videos to research staff. When staff received the date- and time-stamped videos confirming adherence during the appropriate 4-hour dosing window, they replied with text messages that included an electronic statement of voucher earnings (e.g., "Great job! You earned $1 for taking your medication on time today!"). Vouchers, like tokens, function as conditioned reinforcers, which can later be exchanged for retail goods or gift cards (Higgins et al., 2002). Patients received an escalating schedule of vouchers with a reset contingency (Roll and Higgins, 2000) to promote consecutive days of medication adherence. Over the course of the 12-week intervention, patients in the reinforcement condition could earn up to $444 for taking one pill per day or $468 if their medication regimen required twice-daily dosing. Based on pill count assessments conducted in both groups, patients in the reinforcement group were significantly more likely to adhere to their antihypertensive medication (99%) compared to patients in the standard care group (92%).

Summary

Although interventions informed by traditional economic theory can produce reliable increases in medication adherence, the effects of these interventions are typically limited by two major constraints. First, the cost of copayments cannot be reduced to an amount lower than $0. Eliminating copayments may be enough to help patients overcome some financial barriers, but providing additional incentives may be necessary to help patients overcome other barriers to adherence (see Table 7.1). Second, a reduced copayment is most likely to influence behavior precisely when the patient comes into contact with this consequence (e.g., once per month when a prescription is filled); however, medication adherence involves behavior that occurs much more frequently than this (e.g., taking a pill twice per day). Therefore, more frequent programmed consequences may be necessary to modify this behavior.

Reinforcement-based behavioral economics interventions can overcome the limitations of traditional economic interventions by providing positive reinforcers contingent on the occurrence of the targeted behavior (e.g., each time the patient takes a dose of medication as prescribed). The studies of reinforcement-based interventions discussed in this chapter show that these interventions are feasible and efficacious in promoting medication adherence under a variety of conditions (see Table 7.2). They work in diverse populations (e.g., children, adults, and clinical populations) using various reinforcers (e.g., methadone, employment, cash, tokens, prizes, and vouchers) to promote medication adherence in patients with a range of health problems (e.g., alcohol abuse, opioid dependence, schizophrenia, latent tuberculosis, HIV, and high blood pressure).

The results of these studies should be interpreted within the context of two important limitations. First, some of these studies were methodologically flawed. That is, they lacked objective measures of adherence or utilized poor experimental designs (Cass et al., 2005; Liebson et al., 1978; Staring et al., 2010). Second, an exhaustive review of all relevant studies was beyond the scope of this chapter. Studies chosen for discussion were selected because they evaluated relatively unique examples of reinforcement-based medication adherence interventions. Thus, studies that found more modest effects of this type of intervention may not have been discussed. Importantly, however, studies that report minimal or null effects of reinforcement-based interventions on medication adherence are in the minority. Systematic reviews have shown that, overall, these interventions are efficacious (DeFulio and Silverman, 2012; Petry et al., 2012). For example, Petry et al. conducted a meta-analysis of studies that used financial reinforcers to promote medication adherence and found these interventions significantly improved adherence relative to controls across the majority of studies with an overall medium to large effect size of 0.77. Nevertheless, more research on these interventions is needed.

Future research directions

More research is needed to investigate variables that may moderate the effects of reinforcement-based interventions. The efficacy of reinforcement-based interventions is determined, in part, by quantitative and qualitative features of the reinforcers, including reinforcer magnitude, frequency, and delay (see Meredith et al., 2014). Petry et al. (2012) found reinforcement-based medication adherence interventions that utilized greater magnitudes of reinforcement and more frequent reinforcement engendered larger effect sizes. This finding is consistent with results of research in other domains (e.g., Lussier et al., 2006). However, few prospective studies have been conducted to evaluate the effects of these variables on medication adherence. Although one such study conducted by Carroll et al. (2002) found no effect of reinforcer magnitude on medication adherence, this study compared only two different magnitudes of reinforcement.

Much more research is needed on the effects of reinforcement-based interventions on secondary health outcomes. In patients who are moderately adherent to certain medications (e.g., antiretroviral therapy), minor increases in adherence may produce health benefits (e.g., suppression of viral replication).

Thus, major increases in adherence like those observed in most studies discussed in this chapter would certainly be expected to have clinically meaningful effects on health. Although some studies reviewed in this chapter found reinforcement of medication adherence was associated with improvements in biological indices of health (e.g., Rosen et al., 2007), other studies were not powered to detect changes in secondary outcomes (e.g., Petry et al., 2015), or they did not report any health-related outcomes other than adherence (e.g., Cass et al., 2005).

More research is needed on the post-treatment effects of reinforcement-based interventions on medication adherence. Although post-treatment effects are of little concern when a medication regimen is of short duration (e.g., for acute health conditions), they are of major concern when the regimen lasts substantially longer than the intervention (e.g., for chronic health conditions). Yet, only a handful of studies have investigated post-treatment effects of reinforcement-based interventions on adherence, and these studies showed treatment effects diminished over time (e.g., Rosen et al., 2007). Indeed, long-term treatment efficacy is the Achilles' heel of virtually every behavioral, psychosocial, or pharmacological intervention designed to treat a chronic health problem. Treatment effects typically dissipate when interventions are terminated. However, this pattern may be changed within the context of reinforcement-based interventions by adopting affordable strategies to extend treatment duration (e.g., schedule thinning; Preston et al., 2002). Indeed, research suggests increasing treatment duration prolongs treatment effects (Petry et al., 2012).

Research is critically needed to evaluate the cost-effectiveness of reinforcement-based interventions, particularly those designed to promote adherence in patients with chronic diseases. Interventions of considerable duration and, therefore, substantial cost, may be required to treat these patients. Although researchers have developed strategies to minimize or offset the costs associated with reinforcement-based interventions (e.g., Petry, 2000; Silverman et al., 2001), this type of intervention will always come with a price tag. Before implementing reinforcement-based interventions on a wide scale, key stakeholders and policymakers will want to know if the savings in health care costs that result from improved adherence will cover the costs associated with reinforcement-based interventions.

Finally, more research is needed on the acceptability of reinforcement-based interventions to promote medication adherence. Some critics of reinforcement-based interventions are morally opposed to paying patients to do something they believe the patients should be motivated to do on their own (Parke et al., 2013). However, research shows these objections vary considerably depending on the targeted behavior. For example, taxpayers may object more to reinforcement-based interventions that promote smoking cessation compared to those that promote compliance with mental health therapy (Promberger et al., 2011). The extent to which these objections apply to interventions that promote medication adherence is unclear.

Conclusion

The problem of medication nonadherence cannot be overstated. It is a global epidemic. Although many of the solutions that have been proposed to combat this public health threat are complex, the issue itself is as simple as former U.S. Surgeon General C. Everett Koop once put it, "Drugs don't work in patients who don't take them."

This simple problem has a promising, evidence-based solution: reinforcement-based interventions. However, there are research gaps, financial concerns, logistical challenges, and ethical objections associated with these interventions. More research is needed on the long-term effects of reinforcement-based interventions on adherence and health. Data on the cost-effectiveness of these interventions are needed. New procedures and better technology are needed to facilitate remote, objective assessment of adherence. Ethical decisions regarding which patients should benefit from reinforcement-based interventions must also be considered.

Until these issues are addressed, they will likely function as barriers to the large-scale adoption of reinforcement-based interventions to promote medication adherence. However, because researchers, practitioners, and policymakers in other domains have found ways to circumvent many of the same barriers (e.g., Petry et al., 2014), there is reason to remain optimistic about the dissemination potential of reinforcement-based interventions to promote medication adherence.

Just as there is reason for optimism, there is also reason for urgency. As the literature reviewed in this chapter suggests, reinforcement-based interventions are feasible, flexible, and powerful interventions that produce robust improvements in medication adherence. The sooner these interventions can be translated from research into practice, the sooner they can be used to improve the health and well-being of millions of nonadherent patients around the world.

Acknowledgments

Preparation of this chapter was supported in part by National Institutes of Health grants P50-DA09241, P60-AA03510, R01-HD075630, and DP3-DK097705.

References

Bouvy, M.L., Heerdink, E.R., Urquhart, J., Grobbee, D.E., Hoe, A.W., and Leufkens, H.G. (2003). "Effect of a Pharmacist-led Intervention on Diuretic Compliance in Heart Failure Patients: A Randomized Controlled Study," *Journal of Cardiac Failure* 9(5): 404–411.

Brown, M.T. and Bussell, J.K. (2011). "Medication Adherence: WHO Cares?" *Mayo Clinic Proceedings* 86(4): 304–314.

Carroll, K.M., Sinha, R., Nich, C., Babuscio, T., and Rounsaville, B.J. (2002). "Contingency Management to Enhance Naltrexone Treatment of Opioid Dependence: A Randomized Clinical Trial of Reinforcement Magnitude," *Experimental and Clinical Psychopharmacology* 10(1): 54.

Cass, A., Talavera, G., Gresham, L., Moser, K., and Joy, W. (2005). "Structured Behavioral Intervention to Increase Children's Adherence to Treatment for Latent Tuberculosis Infection," *The International Journal of Tuberculosis and Lung Disease* 9(4): 415–420.

Centers for Disease Control and Prevention (2015). "Therapeutic Drug Use." Available at http://www.cdc.gov/nchs/fastats/drug-use-therapeutic.htm (accessed January 29, 2016).

Chernew, M.E., Rosen, A.B., and Fendrick, A.M. (2007). "Value-based Insurance Design," *Health Affairs* 26(2): 195–203.

Cramer, J.A. and Rosenheck, R. (1998). "Compliance with Medication Regimens for Mental and Physical Disorders," *Psychiatric Services* 49(2): 196–201.

DeFulio, A. and Silverman, K. (2012). "The Use of Incentives to Reinforce Medication Adherence," *Preventive Medicine* 55: S86–S94.

DeFulio, A., Everly, J.J., Leoutsakos, J.S., Umbricht, A., Fingerhood, M., Bigelow, G.E., et al. (2012). "Employment-based Reinforcement of Adherence to an FDA Approved Extended Release Formulation of Naltrexone in Opioid-dependent Adults: A Randomized Controlled Trial," *Drug and Alcohol Dependence* 120(1): 48–54.

DiMatteo, M.R. (2004). "Variations in Patients' Adherence to Medical Recommendations: A Quantitative Review of 50 Years of Research," *Medical Care* 42(3): 200–209.

Dunn, K.E., Defulio, A., Everly, J.J., Donlin, W.D., Aklin, W.M., Nuzzo, P.A., et al. (2013). "Employment-based Reinforcement of Adherence to Oral Naltrexone Treatment in Unemployed Injection Drug Users," *Experimental and Clinical Psychopharmacology* 21(1): 74.

Elk, R., Grabowski, J., Rhoades, H., Spiga, R., Schmitz, J., and Jennings, W. (1993). Compliance with Tuberculosis Treatment in Methadone-maintained Patients: Behavioral Interventions, *Journal of Substance Abuse Treatment* 10(4): 371–382.

Farley, J.F., Wansink, D., Lindquist, J.H., Parker, J.C., and Maciejewski, M.L. (2012). "Medication Adherence Changes Following Value-based Insurance Design," *The American Journal of Managed Care* 18(5): 265–274.

Fischer, M.A., Stedman, M.R., Lii, J., Vogeli, C., Shrank, W.H., Brookhart, M.A., et al. (2010). "Primary Medication Non-adherence: Analysis of 195,930 Electronic Prescriptions," *Journal of General Internal Medicine* 25(4): 284–290.

Goldman, D.P., Joyce, G.F., and Zheng, Y. (2007). "Prescription Drug Cost Sharing: Associations with Medication and Medical Utilization and Spending and Health," *JAMA* 298(1): 61–69.

Haynes, R.B. (2001). "Interventions for Helping Patients to Follow Prescriptions for Medications," *Cochrane Database of Systematic Reviews*, 1.

Higgins, S.T., Alessi, S.M., and Dantona, R.L. (2002). "Voucher-based Incentives: A Substance Abuse Treatment Innovation," *Addictive Behaviors* 27(6): 887–910.

Ho, P.M., Magid, D.J., Masoudi, F.A., McClure, D.L., and Rumsfeld, J.S. (2006). "Adherence to Cardioprotective Medications and Mortality among Patients with Diabetes and Ischemic Heart Disease," *BMC Cardiovascular Disorders* 6(1): 48.

Huskamp, H.A., Deverka, P.A., Epstein, A.M., Epstein, R.S., McGuigan, K.A., and Frank, R.G. (2003). "The Effect of Incentive-based Formularies on Prescription-drug Utilization and Spending," *New England Journal of Medicine* 349(23): 2224–2232.

Kazdin, A.E. (1982). "The Token Economy: A Decade Later," *Journal of Applied Behavior Analysis* 15(3): 431–445.

Keller, P.A., Harlam, B., Lowenstein, G., and Volpp, K.G. (2011). "Enhanced Active Choice: A New Method to Motivate Behavior Change," *Journal of Consumer Psychology* 21: 376–383.

Kennedy, J., Tuleu, I., and Mackay, K. (2008). "Unfilled Prescriptions of Medicare Beneficiaries: Prevalence, Reasons, and Types of Medicines Prescribed," *Journal of Managed Care Pharmacy* 14(6): 553–560.

Liebson, I.A., Tommasello, A., and Bigelow, G.E. (1978). "A Behavioral Treatment of Alcoholic Methadone Patients," *Annals of Internal Medicine* 89(3): 342–344.

Lowe, C.J., Raynor, D.K., Purvis, J., Farrin, A., and Hudson, J. (2000). "Effects of a Medicine Review and Education Programme for Older People in General Practice," *British Journal of Clinical Pharmacology* 50(2): 172–175.

Lussier, J.P., Heil, S.H., Mongeon, J.A., Badger, G.J., and Higgins, S.T. (2006). "A Meta⊠analysis of Voucher⊠based Reinforcement Therapy for Substance Use Disorders," *Addiction* 101(2): 192–203.

Maciejewski, M.L., Farley, J.F., Parker, J., and Wansink, D. (2010). "Copayment Reductions Generate Greater Medication Adherence in Targeted Patients," *Health Affairs* 29(11): 2002–2008.

Meredith, S.E., Jarvis, B.P., Raiff, B.R., Rojewski, A.M., Kurti, A., Cassidy, R.N., et al. (2014). "The ABCs of Incentive-based Treatment in Health Care: A Behavior Analytic Framework to Inform Research and Practice," *Psychology Research and Behavior Management* 7: 103–114.

Minozzi, S., Amato, L., Vecchi, S., Davoli, M., Kirchmayer, U., and Verster, A. (2011). "Oral Naltrexone Maintenance Treatment for Opioid Dependence," *Cochrane Database of Systematic Reviews* 16.

Nieuwlaat, R., Wilczynski, N., Navarro, T., Hobson, N., Jeffery, R., Keepanasseril, A., et al. (2014). "Interventions for Enhancing Medication Adherence," *Cochrane Database of Systematic Reviews* 11.

Osterberg, L. and Blaschke, T. (2005). "Adherence to Medication," *New England Journal of Medicine* 353(5): 487–497.

Ottomanelli, G. (1999). "Methadone Patients and Alcohol Abuse," *Journal of Substance Abuse Treatment* 16(2): 113–121.

Parke, H., Ashcroft, R., Brown, R., Marteau, T.M., and Seale, C. (2013). "Financial Incentives to Encourage Healthy Behaviour: An Analysis of UK Media Coverage," *Health Expectations* 16(3): 292–304.

Paterson, D.L., Swindells, S., Mohr, J., Brester, M., Vergis, E.N., Squier, C., et al. (2000). "Adherence to Protease Inhibitor Therapy and Outcomes in Patients with HIV Infection," *Annals of Internal Medicine* 133(1): 21–30.

Petry, N.M. (2000). "A Comprehensive Guide to the Application of Contingency Management Procedures in Clinical Settings," *Drug and Alcohol Dependence* 58(1): 9–25.

Petry, N.M., Alessi, S.M., Byrne, S., and White, W.B. (2015). "Reinforcing Adherence to Antihypertensive Medications," *The Journal of Clinical Hypertension* 17(1): 33–38.

Petry, N.M., DePhilippis, D., Rash, C.J., Drapkin, M., and McKay, J.R. (2014). "Nationwide Dissemination of Contingency Management: The Veterans Administration Initiative," *The American Journal on Addictions* 23(3): 205–210.

Petry, N.M., Rash, C.J., Byrne, S., Ashraf, S., and White, W.B. (2012). "Financial Reinforcers for Improving Medication Adherence: Findings from a Meta-analysis," *The American Journal of Medicine* 125(9): 888–896.

Piette, J.D., Weinberger, M., McPhee, S.J., Mah, C.A., Kraemer, F.B., and Crapo, L.M. (2000). "Do Automated Calls with Nurse Follow-up Improve Self-care and Glycemic Control among Vulnerable Patients with Diabetes?" *The American Journal of Medicine* 108(1): 20–27.

Preston, K.L., Silverman, K., Umbricht, A., DeJesus, A., Montoya, I.D., and Schuster, C.R. (1999). "Improvement in Naltrexone Treatment Compliance with Contingency Management," *Drug and Alcohol Dependence* 54(2): 127–135.

Preston, K.L., Umbricht, A., and Epstein, D.H. (2002). "Abstinence Reinforcement Maintenance Contingency and One-year Follow-up," *Drug and Alcohol Dependence* 67(2): 125–137.

Promberger, M., Brown, R.C., Ashcroft, R.E., and Marteau, T.M. (2011). "Acceptability of Financial Incentives to Improve Health Outcomes in UK and U.S. samples," *Journal of Medical Ethics* 37(11): 682–687.

Roll, J.M. and Higgins, S.T. (2000). "A Within-subject Comparison of Three Different Schedules of Reinforcement of Drug Abstinence Using Cigarette Smoking as an Exemplar," *Drug and Alcohol Dependence* 58(1): 103–109.

Rosen, M.I., Dieckhaus, K., McMahon, T.J., Valdes, B., Petry, N.M., Cramer, J., and Rounsaville, B. (2007). "Improved Adherence with Contingency Management," *AIDS Patient Care and STDs* 21(1): 30–40.

Sabaté, E. (2003). "Adherence to long-term therapies: Evidence for action," World Health Organization.

Seal, K.H., Kral, A.H., Lorvick, J., McNees, A., Gee, L., and Edlin, B.R. (2003). "A Randomized Controlled Trial of Monetary Incentives vs. Outreach to Enhance Adherence to the Hepatitis B Vaccine Series among Injection Drug Users," *Drug and Alcohol Dependence* 71(2): 127–131.

Silverman, K., Svikis, D., Robles, E., Stitzer, M.L., and Bigelow, G.E. (2001). "A Reinforcement-based Therapeutic Workplace for the Treatment of Drug Abuse: Six-month Abstinence Outcomes," *Experimental and Clinical Psychopharmacology* 9: 14–23.

Simpson, S.H., Eurich, D.T., Majumdar, S.R., Padwal, R.S., Tsuyuki, R.T., Varney, J., et al. (2006). "A Meta-analysis of the Association between Adherence to Drug Therapy and Mortality," *BMJ* 333(7557): 15.

Sokol, M.C., McGuigan, K.A., Verbrugge, R.R., and Epstein, R.S. (2005). "Impact of Medication Adherence on Hospitalization Risk and Healthcare Cost," *Medical Care* 43(6): 521–530.

Sorensen, J.L., Haug, N.A., Delucchi, K.L., Gruber, V., Kletter, E., Batki, S.L., et al. (2007). "Voucher Reinforcement Improves Medication Adherence in HIV-positive Methadone Patients: A Randomized Trial," *Drug and Alcohol Dependence* 88(1): 54–63.

Staring, A.B., Mulder, C.L., and Priebe, S. (2010). "Financial Incentives to Improve Adherence to Medication in Five Patients with Schizophrenia in the Netherlands," *Psychopharmacology Bulletin* 43(1): 5–10.

Stitzer, M., Bigelow, G., Lawrence, C., Cohen, J., D'Lugoff, B., and Hawthorne, J. (1977). "Medication Take-home as a Reinforcer in a Methadone Maintenance Program," *Addictive Behaviors* 2(1): 9–14.

The Boston Consulting Group (2003). "The Hidden Epidemic: Finding a Cure for Unfilled Prescriptions and Missed Doses," Available at https://www.bcg.com/documents/file14265.pdf (accessed January 28, 2016).

Tulsky, J.P., Pilote, L., Hahn, J.A., Zolopa, A.J., Burke, M., Chesney, M., and Moss, A.R. (2000). "Adherence to Isoniazid Prophylaxis in the Homeless: A Randomized Controlled Trial," *Archives of Internal Medicine* 160(5): 697–702.

Ward, B.W., Schiller, J.S., and Goodman, R.A. (2014). "Multiple Chronic Conditions among U.S. Adults: A 2012 Update," *Preventing Chronic Disease* 11: E62.

Wei, L., Wang, J., Thompson, P., Wong, S., Struthers, A.D., and MacDonald, T.M. (2002). "Adherence to Statin Treatment and Readmission of Patients after Myocardial Infarction: A Six Year Follow Up Study," *Heart* 88(3): 229–233.

Wright, C. and Moore, R.D. (1990). "Disulfiram Treatment of Alcoholism," *The American Journal of Medicine* 88(6): 647–655.

8

INTEGRATING PRINCIPLES FROM BEHAVIORAL ECONOMICS INTO PATIENT NAVIGATION PROGRAMS TARGETING CANCER SCREENING

Yan Li, Fernando A. Wilson, Roberto Villarreal, and José A. Pagán

Introduction

While advances in cancer screening and treatment technologies have resulted in improved cancer survival rates, substantial disparities in patient outcomes still exist across different demographic and socio-economic groups (Siegel et al., 2014; Smith et al., 2014). Hispanics, a fast growing population that constituted about 17 percent of the United States (US) population in 2010, are particularly vulnerable to many forms of cancer (Siegel et al., 2012). About one out of every three Hispanics will be diagnosed with cancer in their lifetime, and cancer has become the leading cause of death for this ethnic group (American Cancer Society, 2015). Specifically, Hispanic women are more likely to be diagnosed with cervical cancer—and at a more advanced stage—than other major racial/ethnic groups in the US (Siegel et al., 2012). Hispanic men, with a persistently low compliance rate for colorectal cancer screening, are at high risk of colorectal cancer and its associated mortality (Stimpson et al., 2012; Yepes-Rios et al., 2006). These findings highlight the need to identify culturally sensitive, effective interventions and services to increase preventive cancer screening for this at-risk population.

Patient navigation refers to "the support and guidance offered to persons with abnormal cancer screening or a new cancer diagnosis in accessing the cancer care system; overcoming barriers; and facilitating timely, quality care provided in a culturally sensitive manner (Freund et al., 2008, p. 3392)." In addition to focusing on persons with abnormal cancer screening, patient navigation generally covers a wide range of specific services and interventions implemented to improve cancer screening in a wide variety of contexts. Examples of the broad array of services that fall under patient navigation include providing transportation, scheduling appointments with culturally sensitive caregivers, providing interpretation services, and assisting with childcare during scheduled appointments

(Freund et al., 2008). Several randomized controlled trials have shown the effectiveness of patient navigation in improving cancer screening rates, decreasing anxiety associated with screening processes and procedures, and increasing patient satisfaction (Ferrante et al., 2008; Percac-Lima et al., 2009). However, there are still limited studies on how components of navigation interventions take effect and could be further improved.

Behavioral economics provides a rich set of concepts, ideas, and principles that can be used to inform policy and to structure and design high-impact, cost-effective cancer screening interventions by influencing the architecture of choices presented to people that ultimately impact their health and health care utilization behaviors (Purnell et al., 2015). Figure 8.1, adapted from Purnell et al. (2015), summarizes several key principles of behavioral economics as well as their potential applications to improve cancer screening. For example, under the concept of *judgement*, *unrealistic optimism* means that people tend to underestimate their risk of cancer and, thus, this bias implies that personalized cancer risk assessment may be needed for individuals with this trait in order to influence their judgment and better inform their decision-making processes. Under the concept of *decision-making*, *framing effects* may mean that messages about cancer screening need to be tailored to each individual to incorporate personal information and satisfy personal needs. *Social and cultural norms* imply that information and interventions that may influence social norms related to the benefits and risks of cancer screening may help accelerate or decrease the adoption of a given set of behaviors. *Choice architecture* emphasizes *incentives*; for example, providing people (especially low-income people) with low-cost or free screening services and incentives to cover the wide range of economic costs related to undergoing cancer screening (e.g., transportation to screening appointments, time off from work). *Structuring complex choices* means providing detailed explanations of screening choices and alleviating people's concerns with regard to the different choices that they may face at a given point in time.

Although patient navigation programs have proved to be effective in improving health outcomes across different underserved populations, it is unclear whether these successes were attributed to the incorporation of principles of behavioral economics. This chapter discusses the implementation and results from two specific patient navigation programs implemented in San Antonio, Texas, to improve screening for colorectal and cervical cancer for an underserved Hispanic population. We describe the background and components of each patient navigation program, as well as how these programs incorporated principles of behavioral economics to improve screening rates and patient outcomes. We then present results from the cost-effectiveness analysis of both programs and the sensitivity analyses with regard to several key cost and effectiveness parameters. Finally, we conclude the chapter by identifying key potential pitfalls of integrating behavioral economics to patient navigation programs and propose several future research directions.

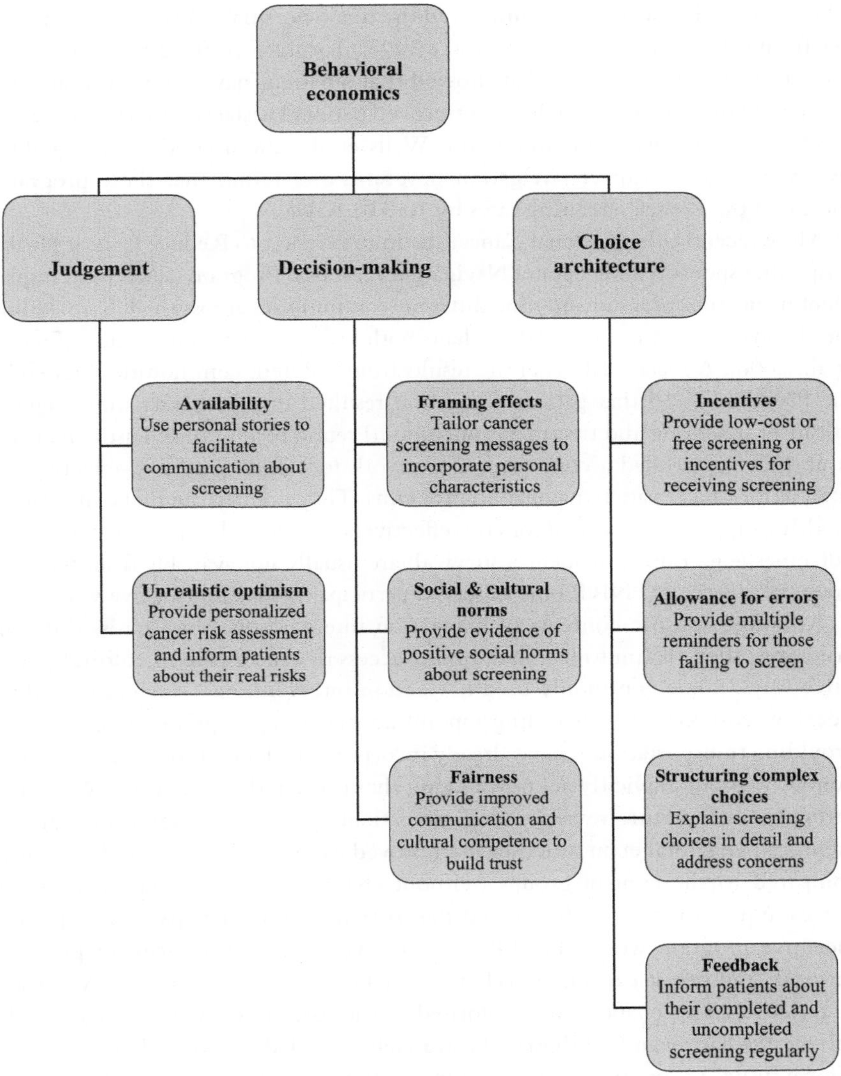

FIGURE 8.1 Concepts and principles of behavioral economics and their potential applications to improve cancer screening.

Source: Adapted from Purnell et al. 2015.

Literature review

Freeman and Rodriguez (2011) presented the history of patient navigation programs in the US. The first patient navigation program was implemented in 1990 in Harlem, New York, to address the disparity of breast cancer screening and treatment in a low-income population (Freeman and Rodriguez, 2011).

The intervention successfully improved the five-year survival rate of breast cancer from 39% to 70% (Oluwole et al., 2003). Ferrante et al. (2008) reported a randomized controlled trial that showed that a patient navigation program implemented in Newark, New Jersey improved patient satisfaction in the follow-up care after an abnormal mammogram. Wells et al. (2008) conducted a qualitative review of 16 cancer navigation programs and found that these programs increased the cancer screening rates by 10.8 to 17.1%.

More recently, the National Cancer Institute's Center to Reduce Cancer Health Disparities sponsored the Patient Navigation Research Program (PNRP) to implement navigation programs in nine different communities across the US, including racial and ethnic minorities and residents with low socio-economic status (Freund et al., 2008). A meta-analysis of the results from different communities funded by the PNRP showed that patient navigation resulted in a moderate improvement in cancer screening and treatment outcomes (Freund et al., 2014). Lastly, Ramsey et al. (2009) provided a conceptual framework for conducting cost-effectiveness analysis for cancer patient navigation programs. They pointed out that a simulation modeling approach is needed for cost-effectiveness analysis because the long-term outcomes (e.g., cancer incidence, survival) are usually not available from the program data (e.g., the PNRP only observed participants over four or five years).

Although specific contents of navigation interventions vary across different programs, they all aim to facilitate timely access to cancer care in culturally sensitive ways. Some commonly used navigation interventions—such as providing free/low-cost services, scheduling appointments with appropriate caregivers, and providing transportation—have already incorporated elements of behavioral economics without explicitly acknowledging them (Freund et al., 2008). We found through our literature review that these patient navigation programs (with elements of behavioral economics) have improved patient satisfaction and outcomes compared to the control groups (without elements of behavioral economics). For example, a recent study showed that patients who participated in a patient incentive program were more likely to receive screening services for prostate, cervical, and breast cancer, as well as other preventive care services (Mehrotra et al., 2014). This chapter aims to formalize the role of behavioral economics in patient navigation and synthesize the available knowledge and evidence to better inform future program designers to improve the cost-effectiveness of the existing patient navigation programs.

Examples of incorporating behavioral economics into cancer screening patient navigation

Navigating Hispanic men to improve colonoscopy screening

Hispanic men have a high risk of colorectal cancer, but a low colonoscopy screening rate compared to other racial/ethnic groups (Yepes-Rios et al., 2006). Recent research has shown that improving colonoscopy screening for Hispanics not only requires improving socio-economic status and access to care, but also necessitates addressing a range of behavioral and cultural barriers such as unrealistic optimism about individual cancer risk as well as fears about the inconvenience and invasiveness of the procedure (Beyer et al., 2011; Freeman, 2006; Jandorf et al., 2013; Shelton et al., 2011; Stimpson et al., 2012). Patient navigation interventions developed based on behavioral economics principles have the potential to help underserved populations overcome these barriers and improve their screening rates.

The patient navigation program we describe below was developed by University Health System (UHS), a public academic medical center located in San Antonio, Texas (Wilson et al., 2014). The target population included Hispanic males, 50 and older, who were members of CareLink (Bexar County's financial assistance program) and who had not received colorectal cancer screening in the last 10 years. Current guidelines recommend regular colonoscopy screenings starting from age 50 (Smith et al., 2014). UHS data showed that only 16 percent of the target population had been screened for colorectal cancer from January 2006 through August 2010 (Wilson et al., 2014). Also, most of the target Hispanic males were low-income heads of households and, as such, being in optimal health was particularly critical for their families and the overall health of the community. Financial barriers (e.g., cost of screening and medication, missing a paycheck due to a screening appointment), transportation inconvenience, and various cultural and social challenges prevented the target population from proactively seeking the recommended screening services.

Consistent with the principles of behavioral economics, the patient navigation program promoted colonoscopy screening services among the targeted Hispanic men by providing a range of culturally sensitive navigation services. The program consisted of four main components: (1) free screening colonoscopy referrals for Hispanic men 50 years of age and older, (2) patient navigation and education from bilingual patient navigators, (3) assisted transportation, and (4) colonoscopy services provided by a bilingual, male Hispanic surgeon.

During the implementation of the program, patient navigators assessed patient needs, informed patients about the colonoscopy process, proper bowel preparation, and conscious sedation, and addressed concerns and fears raised by patients. Patient navigators also provided home visits to further facilitate patients to schedule and receive screening. In addition, patient navigators were equipped with computerized tablets, which facilitated their communication with patients about the risks and consequences of developing colorectal cancer and the benefits of colonoscopy screening, thereby addressing the *unrealistic optimism* of some patients. They also engaged family members of the patient and elicited family support for the screening services. To incorporate the principle of *allowance for errors* (i.e., mistakes are expected and opportunities are given for correction), patient navigators also made calls to remind patients of their missed appointments.

The program also used a patient coordinator to facilitate recruitment, transportation, and social support for patients. Specifically, the patient coordinator conducted community outreach and contracted a livery service to provide assisted transportation to and from the colonoscopy appointment. A livery service was an *incentive* to patients explicitly designed to change behavior and encourage active participation. The patient navigator also provided ongoing cultural and social support consistent with addressing behavioral issues (e.g., *framing effects, social and cultural norms*) that may arise during the period between the referral and the completion of the actual screening procedure. In addition, the patient navigator scheduled and coordinated medical care appointments and set up an electronic reminder system delivered via both personal and automated telephone calls to remind patients of their scheduled appointments.

In addition to the activities performed by patient navigators, the program offered extended clinic hours to provide patients with more flexibility in scheduling their screening during weekend and evening hours. Also, the colonoscopy was free to participants, serving as another *incentive* offered by the program. The program used bilingual medical staff whenever possible, which provided participants with a sense of *fairness* through improved communication and cultural competence. Family members were also well informed of the patient's progress during the procedure to provide additional support. Finally, the program staff monitored and tracked the patient's post-procedure recovery and addressed any concerns to improve patient satisfaction and increase the compliance rate for follow-up screenings.

Cost-effectiveness analysis of the patient navigation program to increase colorectal cancer screening

We conducted the cost-effectiveness analysis of the patient navigation program to improve colonoscopy screening using a stochastic simulation model, based on a widely used modeling framework of the natural history of colorectal cancer (Frazier et al., 2000). Figure 8.2 shows a simplified version of the modeling framework. People with normal mucosa may develop low-risk polyps or advanced adenomas. If screened and detected, these polyps may be removed through polypectomy during colonoscopy. However, if no screening is completed or existing polyps are not detected, they may become malignant, progress to cancer (from localized to distant), and eventually lead to death. Early detection of polyps or cancer is associated with a reduced likelihood of cancer and cancer-related mortality. Clinical guidelines recommend a ten-year screening interval following an initial screening age of 50 for colonoscopy (Smith et al., 2014). People who are diagnosed with cancer through colonoscopy are assumed to begin treatment immediately in the model. People may die either due to colorectal cancer or other causes. We estimated model parameters—including transition probabilities, costs, and quality of life weights—based on the program data and previous studies. See Wilson et al. (2014) for details about parameter estimation.

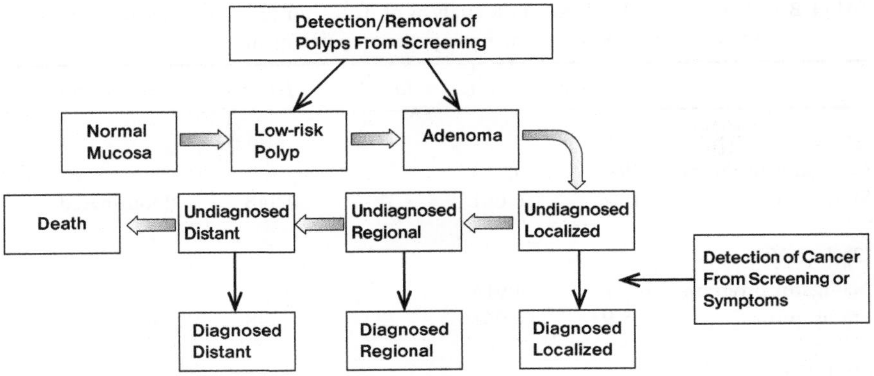

FIGURE 8.2 Conceptual model of colorectal cancer disease progression, detection and screening.

Source: Wilson et al. 2014.

Table 8.1 presents the baseline cost-effectiveness of the navigation program relative to status quo for a 50-year-old Hispanic male program participant. Under a baseline scenario, we showed that the navigation program outperformed status quo for all the health outcomes, including quality-adjusted life-years (QALYs), life-years (LYs), and life expectancy. In particular, we found the navigation program would increase the life expectancy of program participants by six months and result in 0.31 additional QALYs compared to status quo. The program would also reduce the lifetime overall cost of a participant by $1,148. Thus, the program dominated the status quo by resulting in a higher effectiveness and reduced costs.

Figure 8.3 presents sensitivity analyses of the net cost savings of the program relative to status quo if program costs and the effectiveness of the navigator program vary. The figure shows a cost-saving curve that includes the combinations of program cost and effectiveness necessary to maintain a net cost saving of the patient navigation program over status quo. Note that program effectiveness is measured as the percentage of people contacted by the navigator that undergo colonoscopy screening. We found that, with an effectiveness of 80 percent, the program would remain cost saving even if the cost of the navigator program would increase up to 2.5 times (i.e., to $1,250 per person). In addition, under current program costs, the navigation program would remain cost saving as long as the program could maintain a screening rate of at least 18 percent.

Navigating Hispanic women for cervical cancer screening

Despite advances in cancer screening and vaccination in the past few decades, cervical cancer remains a serious threat to population health in the US (Siegel et al., 2013). Significant disparities exist in cervical cancer incidence and mortality by

TABLE 8.1 Predicted cost-effectiveness analysis of navigator program vs. status quo for a 50-year-old Hispanic male program participant

	Outcome	*Cost ($)*	*Incr. Eff.*	*Incr. Cost*	*Cost-effectiveness*
(1) QALY					
Navigator program	19.0	4,913			
Status quo	18.7	6,061	-.31	1,148	Dominated
(2) Life-Years					
Navigator program	19.1	4,913			
Status quo	18.9	6,061	-.22	1,148	Dominated
(3) Life expectancy					
Navigator program	29.9	4,913			
Status quo	29.4	6,061	-.49	1,148	Dominated

Source: Adapted from Wilson et al., 2014.

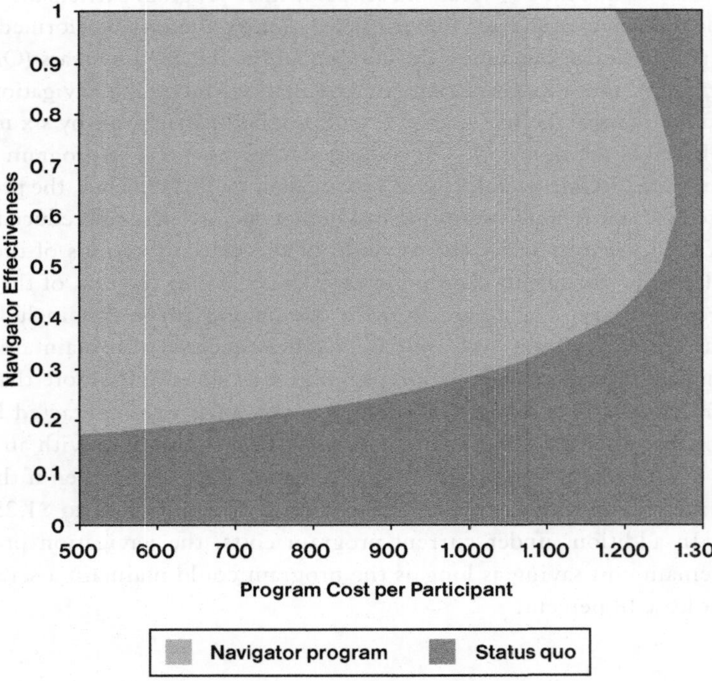

FIGURE 8.3 Sensitivity analysis of net cost savings between navigator program and status quo by program cost per participant and effectiveness of navigator program.

Source: Wilson et al. 2014.

race and ethnicity (Saraiya et al., 2007; Singh et al., 2004). Hispanic women, with a low risk in most forms of cancer, are more likely to be diagnosed with cervical cancer compared to other major ethnic/racial groups (Siegel et al., 2012). The disparity tends to be more significant for vulnerable populations in urban areas (Bastani et al., 2002; Thoms et al., 1995).

We studied a patient navigation program targeting an urban Hispanic population in which 26 percent of women aged 18 and over have not had recommended Pap smear tests within the past three years (Fornos et al., 2014; Li et al., in press). This population has a particularly high risk for cervical cancer, but faces a range of cultural and socio-economic barriers to receiving cancer screening. The patient navigation program is a major component of a community-based, culturally-competent secondary prevention program being implemented by UHS, which is designed to increase the Pap smear test screening rate through the implementation of several key strategies ranging from a mass media health promotion campaign to community outreach and open-access scheduling and patient navigation services (Fornos et al., 2014).

The patient navigation program was designed following principles of behavioral economics. In particular, the program provided personalized social communication by letting participants call "Claudia," a bilingual, female contact person who could be any navigator in the program depending on who was on staff during the day and time in which the contact took place. This is consistent with the principles of *availability* and *social norms* because using the same Hispanic name as a contact person helps participants recall similar events in memory in a culturally competent way. The program also reminded participants of calling "Claudia" with newsletters, public service announcements, and automated client reminders. Afterward, following the lessons from *framing effects* and the principles for *structuring complex choices*, patient navigators provided assessments of the cervical cancer and screening knowledge of patients as well as personalized education about the importance and costs of testing.

In addition to the services provided by patient navigators, the program also implemented multiple-level strategies consistent with principles from behavioral economics. For example, the mass media health promotion campaign allowed women to align their subjective assessment of cervical cancer risk with their actual risk by providing health education and information messages that refer to women similar to or representative of the target population, which addressed issues of *availability* and *social norms*. The program also addressed *unrealistic optimism* by providing accurate information related to cervical cancer risk. Lastly, the program conducted targeted outreach activities (and, thus, appropriately framed issues) rather than conveying general messages to encourage cervical cancer screening. Within three years, the patient navigation program increased the cervical cancer screening rate from 65 to 80 percent among the target population (Li et al., in press).

Cost-effectiveness analysis of patient navigation program to increase cervical cancer screening

We developed a microsimulation model to conduct the cost-effectiveness analysis of the patient navigation program to improve cervical cancer screening. The conceptual framework of the natural history model of human papillomavirus (HPV) infection and the development and progression of cervical cancer was based on existing models (Kulasingam et al., 2011; Myers et al., 2000). Figure 8.4 presents a simplified version of the model structure. We estimated all the parameters based on the program data and existing studies (see Li et al., in press for details about parameter estimation).

Using this agent-based model, we were able to generate 100,000 women with the same age distribution and prevalence of HPV infection as the target population and tracked disease progression for all the simulated individuals throughout their lifetime. Table 8.2 presents the cost-effectiveness results of the patient navigation program compared to status quo using baseline parameters. The cervical program cost an average of $45 more per person than status quo. However, the program would result in an increased life expectancy of 0.2 years and an increase in QALYs of 0.06 years. The incremental cost-effectiveness ratio (ICER) of the navigation program compared to status quo was calculated as $748 per QALY, which means that the program is highly cost-effective (assuming a $50,000 per QALY threshold) under the baseline scenario.

Figure 8.5 presents the results from sensitivity analyses that accounted for uncertainties in the costs and effectiveness of the patient navigation program for cervical cancer. The area under the cost-effectiveness frontier in the figure represented all the possible combinations of program cost per participant and the

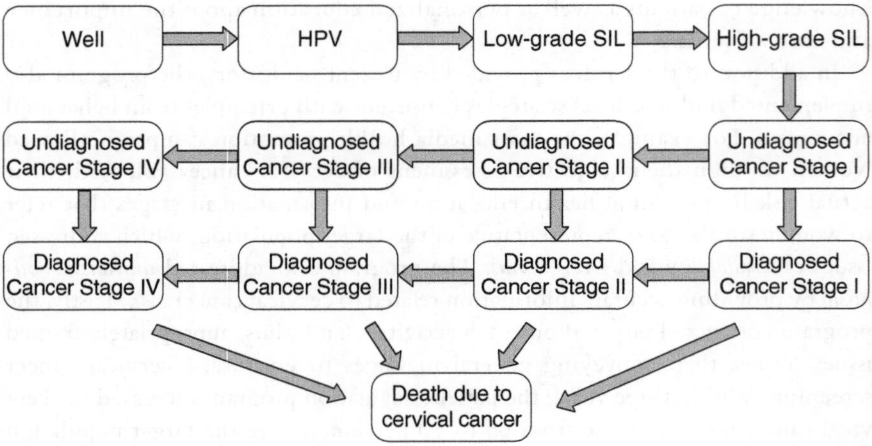

FIGURE 8.4 Simplified model of HPV infection and cervical cancer progression.
Source: Li et al., in press.

TABLE 8.2 Cost-effectiveness of the patient navigation program compared to status quo

Prevention Program	*Cost ($)*	*Life Expectancy*	*QALY*	*ICER ($/QALY)*
Program	642.8	36.49	22.29	748.33
Status quo	597.9	36.29	22.23	
Incr. cost or effectiveness	44.9	0.2	0.06	

Source: Li et al., in press.

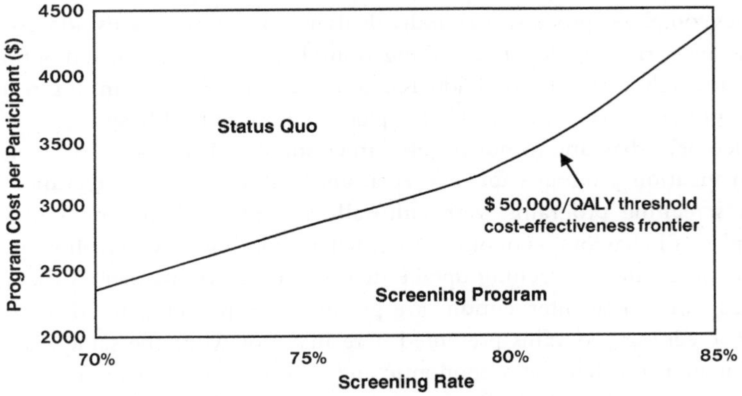

FIGURE 8.5 Two-way sensitivity analysis of the program cost per participant and screening rate for the choice of status quo or screening program given the $50,000/QALY cost-effectiveness threshold.

Source: Li et al., in press.

screening rate that could make the navigation program cost-effective compared to status quo. We found that, under the current program effectiveness (80 percent of program participants receiving Pap screening tests), the cost of the program could increase up to about 10 times (i.e., from $311 to $3,312) before the program crossed the usual cost-effective threshold ($50,000 per QALY). Even if the navigation program was less effective and resulted in a 70 percent screening rate, the program would still be cost-effective as long as its cost remained less than $2,353 per person. Thus, the patient navigation program to improve cervical cancer is highly cost-effective compared to the status quo for the target population. The features and elements of the program represent a well-designed *choice architecture* that can effectively promote cervical cancer prevention for underserved Hispanic women.

Conclusion

Large disparities in cancer screening uptake and outcomes across many socio-economic and demographic groups in the US exist and, despite substantial progress to reduce these differences with the development of new cancer screening initiatives, these gaps in cancer screening and incidence do not seem to disappear (Purnell et al., 2015). Although interventions based on traditional economics—such as the elimination of patient cost-sharing or the adoption of value-based insurance design—may improve cancer screening, they do not explicitly take into account the heterogeneity of the patient population and, thus, may be limited in their ability to reduce health disparities (de Souza et al., 2012; Trivedi et al., 2008). New ideas and approaches to foster behavioral change based on the principles of behavioral economics not only make intuitive sense, but they also show promise in terms of improving uptake and adherence to cancer screening processes and protocols. These new approaches focus on understanding the multiple—and oftentimes complex—processes that individuals use to make decisions, addressing biases in cancer risk and decision-making around cancer screening, and optimizing the choice architecture for individuals considering undergoing cancer screening.

We presented two practical examples on how a health system serving a low-income, urban and minority (Hispanic) population was able to develop patient navigation programs for colorectal and cervical cancer screening. These cancer screening programs were culturally tailored and incorporated several principles of behavioral economics that, when working together, show promise in increasing cancer screening uptake in a cost-effective manner. Although the findings from these interventions are promising, a potential pitfall is that the cancer screening programs presented here may not work the same way when implemented in other socio-economic and demographic groups. For example, an *incentive* in the form of free screening tests may not be as effective in a largely insured population. Another limitation is that it is difficult to disentangle the interrelation among different elements of behavioral economics and assess the influence of each single element in improving the effectiveness of a given patient navigation program. Program designers may have the tendency to integrate as many elements of behavioral economics into their program as possible without fully understanding the roles and interactions of these elements, which may result in a waste of health-related resources. Still, carefully designed cancer screening programs rooted in addressing behavioral biases and change—at least in the cases of colorectal and cervical cancer, as we have shown—seem to delineate a roadmap based on behavioral economics that may be worth pursuing further.

Future research should address how patient navigation programs may work for other forms of cancer (e.g., breast cancer, prostate cancer) and how the results presented above for low income, Hispanic populations in one geographic region of the US may translate to other demographic groups and communities. Moreover, incorporating cancer screening outcomes data for comparison groups is also an important design element that would strengthen the evidence base around what patient navigation programs can do to increase cancer screening rates for different population groups, particularly the most vulnerable. Finally, there is a great potential to expand the power of behavioral economics to facilitate the implementation of screening and treatment programs for other chronic diseases such as hypertension and diabetes.

References

American Cancer Society. (2015). *Cancer Facts & Figures for Hispanics/Latinos 2015–2017.* Atlanta, GA: American Cancer Society. Available at http://www.cancer.org/acs/groups/content/@research/documents/document/acspc-046405.pdf (accessed January 1, 2017).

Bastani, R., Berman, B.A., Belin, T.R., Crane, L.A., Marcus, A.C., Nasseri, K., et al. (2002). "Increasing Cervical Cancer Screening Among Underserved Women in a Large Urban County Health System: Can It Be Done? What Does It Take?" *Medical Care* 40(10): 891–907.

Beyer, K.M., Comstock, S., Seagren, R., and Rushton, G. (2011). "Explaining Place-Based Colorectal Cancer Health Disparities: Evidence from a Rural Context," *Social Science & Medicine* 72(3): 373–382.

De Souza, J.A., Ratain, M.J., and Fendrick, A.M. (2012). "Value-Based Insurance Design: Aligning Incentives, Benefits, and Evidence in Oncology," *Journal of the National Comprehensive Cancer Network* 10(1): 18–23.

Ferrante, J.M., Chen, P.-H., and Kim, S. (2008). "The Effect of Patient Navigation on Time to Diagnosis, Anxiety, and Satisfaction in Urban Minority Women with Abnormal Mammograms: A Randomized Controlled Trial," *Journal of Urban Health* 85(1): 114–124.

Fornos, L.B., Urbansky, K.A., and Villarreal, R. (2014). "Increasing Cervical Cancer Screening for a Multiethnic Population of Women in South Texas," *Journal of Cancer Education* 29(1): 62–68.

Frazier, A.L., Colditz, G.A., Fuchs, C.S., and Kuntz, K.M. (2000). "Cost-Effectiveness of Screening for Colorectal Cancer in the General Population," *JAMA* 284(15): 1954–1961.

Freeman, H.P. (2006). "Patient Navigation: A Community Based Strategy to Reduce Cancer Disparities," *Journal of Urban Health* 83(2): 139–141.

Freeman, H.P. and Rodriguez, R.L. (2011). "History and Principles of Patient Navigation," *Cancer* 117(S15): 3537–3540.

Freund, K.M., Battaglia, T.A., Calhoun, E., Darnell, J.S., Dudley, D.J., Fiscella, K. et al. (2014). "Impact of Patient Navigation on Timely Cancer Care: The Patient Navigation Research Program," *Journal of the National Cancer Institute* 106(6): dju115.

Freund, K.M., Battaglia, T.A., Calhoun, E., Dudley, D.J., Fiscella, K., Paskett, E., et al. (2008). "National Cancer Institute Patient Navigation Research Program," *Cancer* 113(12): 3391–3399.

Jandorf, L., Stossel, L.M., Cooperman, J.L., Graff Zivin, J., Ladabaum, U., Hall, D., et al. (2013). "Cost Analysis of a Patient Navigation System to Increase Screening Colonoscopy Adherence Among Urban Minorities," *Cancer* 119(3): 612–620.

Kulasingam, S.L., Havrilesky, L., Ghebre, R., and Myers, E.R. (2011). "Screening for Cervical Cancer: A Decision Analysis for the US Preventive Services Task Force." Available at http://www.ncbi.nlm.nih.gov/books/NBK92546 (accessed January 1, 2017).

Li, Y., Carlson, E.K., Villarreal, R., Meraz, L., and Pagán, J.A. (in press). "Cost-effectiveness Analysis of a Patient Navigation Program to Improve Cervical Cancer Screening," *American Journal of Managed Care.*

Mehrotra, A., An, R., Patel, D.N., and Sturm, R. (2014). "Impact of a Patient Incentive Program on Receipt of Preventive Care," *The American Journal of Managed Care* 20(6): 494–501.

Myers, E.R., McCrory, D.C., Nanda, K., Bastian, L., and Matchar, D.B. (2000). "Mathematical Model for the Natural History of Human Papillomavirus Infection and Cervical Carcinogenesis," *American Journal of Epidemiology* 151(12): 1158–1171.

Oluwole, S.F., Ali, A.O., Adu, A., Blane, B.P., Barlow, B., Oropeza, R., and Freeman, H.P. (2003). "Impact of a Cancer Screening Program on Breast Cancer Stage at Diagnosis in a Medically Underserved Urban Community," *Journal of the American College of Surgeons* 196(2): 180–188.

Percac-Lima, S., Grant, R.W., Green, A.R., Ashburner, J.M., Gamba, G., Oo, S., et al. (2009). "A Culturally Tailored Navigator Program for Colorectal Cancer Screening in a Community Health Center: A Randomized, Controlled Trial," *Journal of General Internal Medicine* 24(2): 211–217.

Purnell, J.Q., Thompson, T., Kreuter, M.W., and McBride, T.D. (2015). "Behavioral Economics: "Nudging" Underserved Populations to Be Screened for Cancer," *Preventing Chronic Disease* 12. Available at http://www.ncbi.nlm.nih.gov/pmc/articles/PMC4307834/ (accessed January 1, 2017).

Ramsey, S., Whitley, E., Mears, V.W., McKoy, J.M., Everhart, R.M., Caswell, R.J., and et al. (2009). "Evaluating the Cost-Effectiveness of Cancer Patient Navigation Programs: Conceptual and Practical Issues," *Cancer* 115(23): 5394–5403.

Saraiya, M., Ahmed, F., Krishnan, S., Richards, T.B., Unger, E.R., and Lawson, H.W. (2007). "Cervical Cancer Incidence in a Prevaccine Era in the United States, 1998–2002," *Obstetrics & Gynecology* 109(2, Part 1): 360–370.

Shelton, R.C., Jandorf, L., Ellison, J., Villagra, C., and DuHamel, K.N. (2011). "The Influence of Sociocultural Factors on Colonoscopy and FOBT Screening Adherence Among Low-Income Hispanics," *Journal of Health Care for the Poor and Underserved* 22(3): 925.

Siegel, R., Ma, J., Zou, Z., and Jemal, A. (2014). "Cancer Statistics, 2014," *CA: A Cancer Journal for Clinicians* 64(1): 9–29.

Siegel, R., Naishadham, D., and Jemal, A. (2012). "Cancer Statistics for Hispanics/Latinos, 2012," *CA: A Cancer Journal for Clinicians* 62(5): 283–298.

Siegel, R., Naishadham, D., and Jemal, A. (2013). "Cancer Statistics, 2013," *CA: A Cancer Journal for Clinicians* 63(1): 11–30.

Singh, G.K., Miller, B.A., Hankey, B.F., and Edwards, B.K. (2004). "Persistent Area Socioeconomic Disparities in US Incidence of Cervical Cancer, Mortality, Stage, and Survival, 1975–2000," *Cancer* 101(5): 1051–1057.

Smith, R.A., Manassaram-Baptiste, D., Brooks, D., Cokkinides, V., Doroshenk, M., et al. (2014). "Cancer Screening in the United States, 2014: A Review of Current American Cancer Society Guidelines and Current Issues in Cancer Screening," *CA: A Cancer Journal for Clinicians* 64(1): 30–51.

Stimpson, J.P., Pagán, J.A., and Chen, L.-W. (2012). "Reducing Racial and Ethnic Disparities in Colorectal Cancer Screening Is Likely to Require More Than Access to Care," *Health Affairs* 31(12): 2747–2754.

Thoms, W.W., Unger, E.R., Johnson, P.R., Spann, C.O., Hunter, S.H., Smith, R., et al. (1995). "Cervical Cancer Survival in a High Risk Urban Population," *Cancer* 76(12): 2518–2523.

Trivedi, A.N., Rakowski, W., and Ayanian, J.Z. (2008). "Effect of Cost Sharing on Screening Mammography in Medicare Health Plans," *New England Journal of Medicine* 358(4): 375–383.

Wells, K.J., Battaglia, T.A., Dudley, D.J., Garcia, R., Greene, A., Calhoun, E., et al. (2008). "Patient Navigation: State of the Art or Is It Science?" *Cancer* 113(8): 1999–2010.

Wilson, F.A., Villarreal, R., Stimpson, J.P., and Pagán, J.A. (2014). "Cost-effectiveness Analysis of a Colonoscopy Screening Navigator Program Designed for Hispanic Men," *Journal of Cancer Education* 30(2): 260–267.

Yepes-Rios, M., Reimann, J.O., Talavera, A.C., de Esparza, A.R., and Talavera, G.A. (2006). "Colorectal Cancer Screening Among Mexican Americans at a Community Clinic," *American Journal of Preventive Medicine* 30(3): 204–210.

9

BEHAVIORAL ECONOMICS AND HIV

A review of existing studies and potential future research areas

Sebastian Linnemayr

Introduction

In the United States, over 1.2 million people are living with the human immunodeficiency virus (HIV), and almost 150,000 of them (or more than 1 in 8 infected persons) are not aware that they are carrying the virus (Centers for Disease Control and Prevention, 2015b). While the overall incidence (rate of new infections) has remained stable at about 50,000 new infections per year in the US, some groups such as men who have sex with men (MSM) and transgender women (TGW) have difficulty successfully carrying out preventive behaviors, and carry a disproportionate disease burden: MSM make up approximately 2 percent of the US population, but more than half (57%) of all people living with HIV in the US in 2011 were MSM, and 63 percent of all new HIV infections in 2010 occurred among MSM (Centers for Disease Control and Prevention, 2015a). HIV especially affects MSM in urban settings in the US, with some areas having prevalence rates as high as those found in sub-Saharan Africa (Perneger, 1999), where, as of 2013, close to 25 million people were living with the disease (AVERT, 2015).

As in more developed countries, antiretroviral treatment (ART) has improved the life expectancy of HIV-infected patients in many countries in sub-Saharan Africa dramatically (Kebba et al., 2002; Mermin et al., 2004; Mermin et al., 2008). However, the success of ART is dependent on high medication adherence and continued treatment, which are needed to achieve optimal clinical outcomes such as slowing the progression to AIDS, lengthening survival, sustained viral suppression, and prevention of drug resistance and loss of treatment options (Abaasa et al., 2008; Bangsberg et al., 2006; Garcia de Olalla et al., 2002; Hawkins and Murphy, 2007; Nachega et al., 2010; Shuter, 2008).

In the US and abroad the HIV epidemic continues to pose a global health threat—over 30 years after the virus was identified. HIV is a chronic disease, and requires a two-pronged approach: on the one hand, there is a need to support preventive behaviors such as (female) condom use, pre-exposure prophylaxis (PrEP), microbicides, or male circumcision for those at risk of contracting HIV, but it is also necessary to successfully link those who have contracted the virus to appropriate medical care to achieve viral suppression. This is particularly important in light of the findings of the HPTN052 study that showed the potential of "treatment as prevention," that is, the insight that consistent, high ART adherence is protective of HIV transmission. Given the large pool of people who are carriers of the virus, successful linkage to care and high adherence to antiretroviral medication are key priorities on the international public health agenda. Reflecting this thinking has been a reduction in the time between HIV detection and start with ART, both in the US (where treatment initiation is recommended irrespective of CD4 count or disease stage; World Health Organization, 2015) as well as internationally, where the guidelines for ART start have been repeatedly revised to start increasingly earlier after HIV testing (Kongings et al., 2012).

At the same time, ART medication is changing and now can often be taken as a once-daily combination pill. While these newer, simpler ART regimens are more forgiving (Kobin and Sheth, 2011; Parienti et al., 2010; Pasternak et al., 2012), lead to fewer side effects, and have to be taken less often, they result in only marginally better adherence (Nachega et al., 2014). Adherence remains key to the success of ART (Foca et al., 2014; Genberg et al., 2012; Li et al., 2012), and at least 80–85 percent adherence is needed to sustain viral suppression and prevent drug resistance (Kobin and Sheth, 2011; Parienti et al., 2010; Pasternak et al., 2012). Despite the importance of adherence and simpler regimens, mean ART dose-taking adherence (% of prescribed doses taken) typically ranges from 60–80 percent when measured electronically, and only 30–60 percent of patients achieve 85 percent adherence (Gardner et al., 2011; Ortego et al., 2011; Shuter et al., 2012). These data highlight the fact that many people struggle to adequately adhere to ART, even with the newer, simpler regimens. Adherence interventions are particularly important in sub-Saharan Africa where alternative regimens are often prohibitively expensive or altogether unavailable if people develop drug resistance as a result of suboptimal adherence, yet few sustainable interventions exist (Gray et al., 2011; Simoni et al., 2008).

A useful way to conceptualize the behavioral steps from HIV infection to viral suppression—an undetectable level of HIV in the body that is the goal of HIV care—for the rest of this chapter is the HIV care continuum, sometimes also called the HIV treatment cascade (AIDS.gov, 2015). The concept of the treatment cascade was popularized by Edward Gardner and colleagues in a landmark 2011 article in the journal *Clinical Infectious Diseases* and has since been adopted by many agencies at the international, federal, and local levels. The cascade is comprised of five steps (diagnosis of HIV infection, linkage to care, retention in care, receipt of antiretroviral therapy, and achievement of viral suppression) that allow identifying gaps in HIV services (i.e., steps in which a large drop-off of patients is observed) in order to improve strategies to facilitate engagement in care and support healthy outcomes for people living with HIV.

Efforts at improving preventive behaviors so far have shown limited success: following the largely failed campaigns to promote sexual abstinence and the disappointing rates of consistent condom usage even in high-prevalence situations such as among youth in Uganda (IRINnews, 2012), new methods of HIV prevention have been investigated and promoted but have to date not produced the desired results. For example, after randomized trials showed that male circumcision can effectively decrease the risk of HIV acquisition by over 50 percent, international organizations such as the World Health Organization and UNAIDS suggested scaling-up of this procedure in Eastern and Southern Africa (World Health Organization and UNAIDS, 2007). Despite substantial investments by the international community and the concerned countries, a majority of the target countries have not seen the recommended circumcision prevalence rate of 80 percent (Sgaier et al., 2014).

HIV testing is necessary for a person to know they are infected, and represents the first stage to accessing treatment and care that is needed for a person to stay healthy. In the US about 86 percent of people living with HIV are aware of their status, which has led to calls to improve our efforts to get more people at risk to test for HIV (Centers for Disease Control and Prevention, 2014). The next steps in the treatment cascade paint a sobering picture: of the 1.2 million people living with HIV in the US, only about 40 percent are estimated to be engaged in HIV medical care, 37 percent are prescribed antiretroviral treatment (ART), and only 30 percent have successfully achieved an undetectable viral load. This means that 7 out of 10 people living with HIV do not have their virus under control and are at risk of developing opportunistic infections and subsequently succumbing to AIDS in the absence of treatment. Finding novel interventions to improve each of the steps in the treatment cascade is therefore of primary importance. Both the continued high prevalence and incidence of HIV, as well as failure to detect HIV infections and link those concerned effectively to HIV care, suggest that novel strategies are needed to get the HIV epidemic under control. In this chapter, we lay out which strategies based on traditional economics have been targeted at each of the behaviors in the treatment cascade, and to what extent they have been successful. Subsequently, we discuss from a behavioral economics perspective which biases may prevent more successful outcomes, how they could be overcome, and what interventions based on these insights to date have been tried, and whether they were successful. We conclude with a discussion of several potential fruitful avenues for behavioral economics-based research to improve HIV-related behaviors, but also point out potential limitations and ethical considerations that should be taken into account when trying to implement these novel interventions.

Traditional economic approaches

HIV prevention

Traditional economics views the demand for sex (that is the major—though not only—determinant of HIV infections in the US and many sub-Saharan African countries) as guided by the same factors as those influencing any other behavior: costs and benefits. The demand for (unprotected) sex increases when costs (including the risk of contracting sexually transmitted diseases including HIV, but also actual price paid in commercial transactions) are lower. These predictions (to a certain extent) are played out in a study investigating the supply of sex among sex workers in Kenya (Manda, 2013). The price of unprotected sex rises when the risk of a sexually transmitted disease, including HIV transmission, increases, as well as when the risk of pregnancy increases for the sex worker. From this perspective, economic incentives serve foremost to increase the opportunity cost of unprotected sex, such as when an intervention makes payments conditional on remaining HIV-negative or staying free of sexually transmitted diseases, which in turn makes it more costly to risk the possibility of an infection due to an unprotected sex act. To the extent that unprotected sex is the result of a lack of economic agency, empowerment, or credit constraints, financial payments to vulnerable populations can also serve to improve the economic position or outlook of the target participants, and in that way indirectly lead to a reduction in the supply of unprotected sex.

Financial incentives have been tested to incentivize safe sex and prevent HIV and other sexually transmitted infections (STIs) (Kohler and Thornton, 2011; Baird et al., 2012; de Walque et al., 2012; Galárraga et al., 2013). Pettifor et al. (2012) conducted a comprehensive literature review to identify studies providing cash incentives aimed at HIV prevention, and correspondingly classified the studies found into two main categories: those providing cash transfers (either conditional on the absence of evidence of certain sexual risk behaviors and associated outcomes), and those aimed at removing economic barriers, in particular schooling that is often thought to be protective for women for a range of health-related outcomes (Jukes et al., 2008; The World Bank, 2002). The authors found that the majority of studies reviewed that focused on reducing risky sexual behaviors have desired effects, but point out that there is a dearth of published studies that report directly on HIV outcomes. They note that it is not clear what the most effective way of implementing interventions using cash payments is, or whether such interventions have the potential to reduce HIV infection across populations. They also note that a common concern of such studies is their sustainability, given the relatively large payments provided.

Two recent large multisite international trials used cash payments with the goal of reducing HIV infections and have received considerable attention: HPTN 068 was conducted in South Africa and provided participating families in the intervention condition with a monthly payment of about $24 conditional on the child attending school at least 80 percent of the time up until the twelfth grade. However, at the end of the study, there was no significant difference in the incidence of HIV infection between the intervention and the control group, and no difference was found regarding the girls' age of sexual initiation or the number of paid sex actions. The second study, called CAPRISA 007, tested the impact of cash incentives on HIV infection conditional on several conditions related to academics, as well as undergoing tests for HIV, herpes simplex virus 2 infection, and HIV infection. The conditional payments provided to girls in high school were considerable: participants could receive up to $150 a month. While at the end of the study, rates of herpes infections compared to those in the control group were about 30 percent lower, there was no effect found on the number of HIV infections over the course of the study. The study also found no impact on academic performance (Lubinski, 2015).

The studies discussed so far suggest at best a mixed impact of providing (traditional) cash incentives to reduce HIV infection and other related outcomes. There may also be negative consequences of providing relatively large cash incentives: a study by Kohler and Thornton (2011) did not find an effect on HIV status or reported sexual behavior, but reported that male participants who had received a cash transfer were 9 percentage points *more* likely to engage in risky sex, countering the objectives of the original study. This increase in sex could constitute an income effect that subsequently led to an increase in the demand for sex (as for any normal good, having more income means demanding more of those goods). For women, on the other hand, receipt of a cash transfer led to a reduction in risky sexual behavior, potentially indicating a protective effect of increased cash availability through empowerment or the alleviation of credit constraints.

HIV testing

Just as for sexual behaviors, the demand for HIV testing can be modeled from a traditional economics perspective to be either driven by structural barriers, or direct cost/benefit considerations of the demand for learning one's HIV status. A study by Thornton (2008) investigated the demand for HIV testing in an innovative way by randomly varying the geographic location of HIV testing centers, thereby increasing the transport cost of obtaining the results, as well as varying the incentive payments provided for picking up the test results (Thornton, 2008). The evidence suggests that cash incentives have a strong positive impact on the picking up of HIV test results: incentives worth less than a tenth of a day's wage doubled the rate of result-seeking among respondents; and greater distance to the test center, which increases the costs of learning one's results, had a strong negative impact on learning the test results.

One issue to note is that HIV testing is not a one-off behavior, but is recommended at frequent intervals for individuals at risk and, therefore, poses similar problems in terms of sustainability as that of any other repeated chronic behavior. Furthermore, as suggested below, economic/structural barriers may be only a relatively minor issue in the complex decision to learn one's HIV status, as even in resource-poor countries HIV testing is commonly offered free of charge, in convenient locations, and at frequent intervals, yet not enough people attend these testing sessions and are hence unaware of their HIV status. Given the relatively small cash incentives provided, the rewards offered in the Thornton study therefore may actually act to counter present bias or psychological barriers to learning one's test results, rather than addressing structural barriers; as discussed below, it is not always straightforward to classify studies as being of a traditional economics or behavioral economics flavor as the rewards offered may address both structural issues (i.e., have price or income effects), but at the same time also address psychological biases.

Linkage to care

There have been relatively few studies investigating the use of economic incentives to influence the decision of individuals to access (and remain in) care once learning of their HIV infection: an exception is the large-scale HPTN 065 study, also known as the Test, Link-to-Care Plus Treat (TLC-Plus) study. Over the course of three years, HPTN 065 investigated the feasibility of a community-focused strategy to expand HIV testing, diagnose HIV infection, link HIV-positive individuals to medical care, initiate treatment according to current guidelines, and ensure that patients reach viral suppression by adhering to their treatment regimens, that is, follow the major steps in the treatment cascade outlined above (HPTN, 2015). Those in the intervention received a coupon that they could redeem for a $25 gift card. If they subsequently met with an HIV provider to learn of their test results, they would then also receive a $100 gift card. The results of the study were mixed: while linkage-to-care was improved for most of the testing sites participating in the study irrespective of intervention assignment, there was no significant improvement in the proportion of persons seeking medical care after testing HIV-positive, or in the fraction of people with an undetectable viral load, despite further gift cards worth $70 conditional on this outcome.

ART adherence

The use of cash payments to improve medication adherence generally (as laid out in this book in the chapter by Meredith and Petry) as well as to antiretroviral medication (ART) specifically, has a long and varied history. From a traditional economics perspective, such payments alter the price of taking one's medication, either by defraying the costs of the medication taken and/or by increasing the cost of non-adherence. In general, traditional economics has trouble explaining why an individual would not engage in a behavior that in the case of

ART adherence is ostensibly crucial to his survival, and has to assume preferences that put little value on the individual's longer-term health outcomes. An alternative explanation from this viewpoint is a high financial cost of adherence, which is problematic given that in many situations people (even in sub-Saharan Africa) have access to free medication. As discussed above for the case of cash payments for HIV prevention behaviors, cash payments can either have a price effect by making it more expensive to not take one's medication, or increase the person's economic opportunities, that is, have an income effect.

A recent review on the use of conditional economic incentives for improving ART adherence discusses four randomized controlled trials, three of which were targeted at individuals with substance use issues (Galárraga et al., 2013). The trials all showed limited study durations and sample sizes, and provide evidence that conditional incentives can work to improve adherence. However, adherence is found to tend to revert to baseline levels once the incentives are withdrawn. Furthermore, the authors note that there is little insight as to why, when, and for whom incentives work, and that incentivization over longer time periods (as is required for life-long ART adherence) is likely problematic, in particular if relatively large monetary incentives are provided.

Behavioral economics approaches

We noted above that traditional economics has difficulty explaining why people engage in unhealthy behaviors that run counter to their own interests. In this section, we argue that behavioral economics offers a richer perspective on the unhealthy behaviors of engagement in risky sex, failure to get tested for HIV, lack of engaging with available HIV care, or lack of viral suppression despite the relatively wide availability of ART for many people living with HIV.[1] Behavioral economics can shed light on the behaviors along the treatment cascade, many of which share similarities with other chronic health behaviors discussed elsewhere in this volume (Linnemayr, 2015). In this section, we review behavioral economics interventions that have addressed some of these behaviors, and point out behavioral biases that are likely to exert a detrimental influence but that also offer the opportunity to be used as entry points for interventions.

We would like to stress that it is not always possible to draw a clear line between "traditional" and "behavioral" economics approaches underlying programs using monetary incentives to influence health behaviors, and that to some extent our classification of interventions as either traditional or behavioral economics-based is somewhat arbitrary. What is clear is that conditional cash transfer programs as traditionally used typically provide relatively large financial rewards in order to remove structural barriers such as transport costs, to reduce overall poverty in a target population (and are hence likely less effective in well-to-do populations), and/or to override a revealed preference of an individual through changes in price. For the purpose of this chapter, we denote behavioral economics approaches as those using relatively small rewards—those that are unlikely to change prices or lead to income effects—compared to these traditional approaches, as well as interventions that explicitly target specific behavioral biases, and/or use incentives that are designed based on behavioral economics principles.

Prevention

As already discussed above, there is a range of biomedical tools at the disposal of people who perceive themselves to be at risk of HIV infection including male and female condoms, pre-exposure prophylaxis—or PrEP—that consists of taking of ART by HIV-negative people to prevent the transmission of the virus, microbicides, and male circumcision. However, in both clinical trials and open-label use, take-up of and adherence to these prevention methods often fall below protective levels. Linnemayr (2015) points out three key behavioral biases that may contribute to this situation: first, the *salience* of HIV and its consequences for people not infected with the virus may be low as HIV is largely a disease that often cannot be inferred from an infected person's appearance. Therefore, it is easier to push aside precautionary ideas during a sexual encounter as the sexual partner who may be HIV-positive is likely not to show any disease symptoms (compare this scenario to the effort people make, for example, to avoid a fellow commuter with the flu who is loudly coughing on a crowded subway train). Contributing to the low salience of HIV may be its chronic, long-term nature, which results in even initially highly motivated people over time forgetting about the importance of HIV prevention as more pressing concerns of daily life take over a person's mental list of priorities. Lack of salience of HIV and other chronic conditions casts doubt on one-off interventions, and suggests forms of interventions that are repeated, and ideally specific to a person's situation to avoid boredom from being exposed to repetitive messages. As mentioned above, it is likely not lack of information per se ("HIV is a threat to your health") that is a barrier to health HIV-related behaviors, but the challenge of keeping a person focused on the disease risk over time ("Today is Friday; if you go out to a bar and meet someone remain safe").

The second bias likely countering HIV prevention behaviors discussed in Linnemayr (2015) is *present bias* that is likely particularly pronounced for a disease that in many instances in the US and sub-Saharan Africa is transmitted through sexual intercourse. Sexual arousal has been shown to detrimentally influence decision-making (Loewenstein, 2005), and compounds existing self-control problems for HIV prevention where the short-term temptation of carefree, unprotected sex often trumps considerations of potential negative subsequent health consequences in the distant future. The likely importance of present bias in repeated interactions points to the need to think carefully about how incentives to prevent HIV transmission are designed (Linnemayr and Rice, 2016). Again, small and repeated "nudge"-type incentives may prove to be more effective at altering behavior at the right moment than providing relatively large one-off payments conditional on an HIV test. For HIV prevention and other chronic health behaviors, another area of research deserving more attention is how to best reconcile the conflict between external incentives and intrinsic motivation that has sometimes been cited as a criticism of economic incentives. Rather than thinking about these as mutually exclusive, we should think about ways to provide the right type and amount of external incentives to support a person's intrinsic motivation that is necessary to keep up healthy daily behaviors for chronic conditions.

A noteworthy recent study that draws on some of these insights to support HIV prevention among men and women in Lesotho is (Nyqvist et al., 2015). The authors gave out lottery tickets to a high-value lottery ticket group and a low-value ticket group conditional on negative tests for STI screenings every four months during the two-year study; a control arm did not receive such tickets. Lottery drawings were conducted every four months, with two men and two women out of about 50 to 70 local participants in each village winning, with prizes of either about $100 for the high-value group and $50 for the low-value group. These lottery incentives were found to reduce the HIV incidence rate by 2.5 percentage points, or 21.4 percent, resulting in a 3.4 percentage points lower HIV prevalence rate in the pooled intervention group compared to the control group, despite the low expected value of the prizes paid out per participant.

Lotteries offer several advantages over conventional conditional cash transfer (CCT) programs: first, while the value of the lottery prize is higher than the typical CCT amount, the overall payout is lower, supporting sustainability. Second, to the extent that sexual risk taking and monetary risk taking are correlated, risk-prone individuals may be particularly attracted by lotteries compared to fixed payments, thereby targeting the appropriate population segment. Lotteries also leverage the commonly documented behavioral economics bias of overestimation of small probabilities: people tend to overestimate small percentages, and therefore prefer a small chance at a large reward to a small, certain reward (Kahneman and Tversky, 1979; Kahneman, 2011; Barberis, 2013). The return from participating in this form of gambling is hence higher than the return from an incentive program that pays the expected return with certainty from the perspective of the participant, and lotteries therefore may provide stronger incentives for behavioral change compared to a traditional CCT (holding the budget constant).

Male circumcision

A recent study in Kenya tested the impact of providing different types of incentives to improve take-up of voluntary male circumcision in a randomized controlled trial with three groups—a control group, a group receiving conventional fixed cash payments worth approximately $12.50 for undergoing the procedure, and one group that was eligible for lottery incentives conditional on getting circumcised. The lottery prizes were valued from a $2.50 food voucher to larger prizes such as a bicycle or smartphone. The authors found that very few participants in the control group (just over 1%) underwent the procedure within three months, compared to over 8 percent of those in the group receiving fixed payments that may have reduced economic barriers to the procedure such as the opportunity cost of missing work, and may have countered present bias. Interestingly, those in the lottery group showed lower rates of circumcision than those in the fixed-payment group (Witherspoon, 2015). The study underlines that we still lack a clear understanding of how to best implement BE-based interventions, and in which situations and for which target groups they may be most effective.

Adherence

Adherence to antiretroviral treatment (ART) requires life-long, (twice) daily, consistent pill-taking, putting habit-formation and supporting intrinsic motivation in the forefront, as it is difficult to provide external rewards over a long time period in a way that is both sustainable and attractive to the person. Recent evidence indicates that clients take "drug holidays" from their treatment when they feel overwhelmed by the daily task of life-long medication adherence. These gaps in adherence dramatically reduce the effectiveness of ART medication, and highlight the importance of motivation and self-control in fighting treatment fatigue (Nakiyemba et al., 2006). From a behavioral economics perspective, the daily effort to take one's medication may deplete mental capacity (Mani et al., 2013), leading people to increasingly give in to short-term temptation at the expense of long-term benefits (Volpp et al., 2009). Present bias may then prove particularly damaging for chronic conditions such as HIV/AIDS where the benefits of a healthier and longer life occur in the distant future, but the costs of consistently adhering to ART are incurred daily (Linnemayr, 2015).

Two recent papers provide empirical evidence for the importance of present bias as a barrier to ART adherence. (Thirumurthy et al., 2015) present a study among HIV-infected adults initiating ART who at baseline were asked about their time preferences. This, in turn was then used to classify them as having either a low or a high discount rate for the future, indicating that they are more impatient as they perceive relatively little utility from future benefits. They find that 44 percent of participants reported high discount rates. While adherence after 48 weeks did not differ significantly between these two categories of study participants, those with higher discount rates were found to have significantly higher mortality (9.3% vs. 3.1% of those in the "low" discount rate category). In Linnemayr and Stecher (2015), the authors investigate the prevalence of the behavioral economics biases of present bias, overoptimism, and lack of salience among a sample of 155 treatment-mature HIV clients in Uganda and find that 36 percent are classified as present-biased, 21 percent as overoptimistic, and only 34 percent as having salient HIV information (Linnemayr and Stecher, 2015). These biases also seemed to be correlated with adherence over the subsequent four months: present-biased patients showed a 13 percentage-points lower probability of displaying adherence rates above 90 percent, overoptimistic clients were 9 percentage-points less likely, and those not having salient HIV information were 17 percentage-points less likely to show mean adherence of 90 percent or more. The findings from these two papers indicate that behavioral economics could be a useful tool to be used to screen patients in HIV care for likely future adherence problems and to better design and target interventions addressing these behavioral biases and the suboptimal adherence associated with them.

A currently ongoing NIH-funded R34 intervention by the chapter's author called "Rewarding Adherence Program" (RAP) (R34MH096609, clinicaltrials.gov identifier NCT02503072) incentivized ART clients in Uganda to remain engaged in care and highly adherent to their medication. In a 20-months' long randomized controlled trial (RCT), the 155 treatment-mature clients who were asked about their present bias and other behavioral biases discussed above were assigned to either a control group receiving usual care or one of two treatment groups: the first treatment group was eligible for prize drawings each time they came to the clinic and kept their scheduled appointment. Clients in the second treatment group were eligible to draw a prize each time they came for a clinic visit and showed at least 95 percent adherence as measured by medical event monitoring system (MEMS)-caps, which are electronic pill caps recording the date and time a pill box is opened. ART adherence of 90 percent or higher was 23.7 percentage-points more likely in the combined intervention groups than in the control group, suggesting that small prizes allocated by a drawing may be effective at improving adherence (Linnemayr et al., 2017).

Conclusion

In this chapter, we discussed the use of behavioral economics insights to improve HIV-related behaviors along the treatment cascade, ranging from prevention to adherence to antiretroviral medication and viral suppression. We also discussed the use of traditional economics incentives, in particular conditional cash payments, and found that there is mixed evidence as to their effectiveness, little insight as to why they work when they work, and noted that for many studies behaviors seem to go back to baseline levels after the incentives are withdrawn.

Behavioral economics may provide an approach to address some of these unhealthy behaviors that account for the continued spread of the HIV epidemic in a more nuanced way. Unfortunately, currently few studies exist that make use of this rich toolset, and more research is needed to arrive at a better understanding of how to best design and implement such interventions.

A particularly promising area of study is the implementation of behavioral economics incentives using mobile health (mHealth) approaches. Most HIV clients in the US, and in particular those who have been on ART for a number of years, do not see their provider more than once every three months, and it is difficult to support clients between visits that way. Consequently it is important to find other ways of remaining in contact with and supporting people with chronic, life-long conditions during the "5,000 hours a year they do not see a provider" (Asch et al., 2012). mHealth interventions show great promise because of an ever-increasing prevalence of electronic devices, and have been shown to facilitate alerts and reminders, data collection, adherence measurement, direct voice communication, and educational messaging in HIV populations (Catalani et al., 2013). Behavioral economics suggests that small "nudges" frequently provided can address biases and effectively alter behavior, and mHealth provides a way of doing so (for example, by sending participants nudges or incentives to their cell or smart phones). Rather than focus on the magnitude of the prize, behavioral economics suggests that the *way prizes are given out*—and at what time intervals—determines their effectiveness. Provision of such incentives frequently using mHealth may provide a promising way to reach an ever-increasing number of HIV clients in life-long care. Similar considerations apply to HIV prevention that is equally a chronic and long-term behavior that may benefit from the frequent provision of reminders to increase salience and incentives to counter present bias.

As pointed out above, another area of research deserving further attention is the use of Behavioral economics nudges to support chronic health behaviors over the longer run. In particular, we need to learn more about how people's intrinsic motivation—that is arguably crucial for chronic, life-long HIV-related behaviors—can be better incorporated into and supported by interventions providing external incentives. Traditional economics-based payments can typically not be sustained for longer periods of time, and more nuanced design of incentives may hopefully also reduce the often observed reversion of incentivized behaviors to baseline levels once incentives are withdrawn.

Ethical concerns need to be adequately addressed when using economic incentives to influence health behaviors, whether inspired by traditional or behavioral economics (Halpern et al., 2009). However, the incentive amounts used in behavioral economics interventions tend to be small, can be provided in-kind (such as in the chapter author's RAP intervention discussed above), or can make use of non-monetary incentives such as when using social norms that leverage the tendency of many people to take note of the behavior of their peers. All these facts make it less likely that the incentives used are seen as a coercive "payment" for appropriate adherence or prevention behavior. Furthermore, given the positive impact of prevention as well as of adherence on other people's health through the reduced likelihood of HIV transmission through participants showing high adherence, offering incentives is also more defensible than for behaviors that do not show such positive spillovers. Small incentives targeting specific health behaviors—in contrast to contingent cash payments that are typically targeted at those fulfilling certain poverty criteria—have the additional advantage that they can work in any socio-economic environment because they target universal behavioral biases—biases that are likely to be found irrespective of a person's wealth status such as present bias.

Note

1 This is not to reduce the importance of making more readily available these life-saving drugs for millions of people in need in resource-limited settings, as well as many people lacking access in relatively well-to-do countries such as the US for lack of insurance and other reasons that are important policy issues to address in their own right.

References

Abaasa, A.M., Todd, J., Ekoru, K., Kalyango, J.N., Levin, J., Odeke, E., et al. (2008). "Good Adherence to HAART and Improved Survival in a Community HIV/AIDS Treatment and Care Programme: The Experience of The AIDS Support Organization (TASO), Kampala, Uganda," *BMC Health Services Resesearch* 8: 241.

AIDS.gov. (2015). "What Is the HIV Care Continuum?" *Policies/Issues.* Available at https://www.aids.gov/federal-resources/policies/care-continuum/ (accessed January 16, 2017).

Asch, D.A., Muller, R.W., and Volpp, K.G. (2012). "Automated Hovering in Health Care—Watching Over the 5000 Hours," *New England Journal of Medicine* 367(1): 1–3.

AVERT. (2015). "HIV and AIDS in Sub-Saharan Africa Regional Overview," *Regional Overview.* Available at http://www.avert.org/professionals/hiv-around-world/sub-saharan-africa/overview (accessed January 16, 2017).

Baird, S.J., Garfein, R.S., McIntosh, C.T., and Ozler, B. (2012). "Effect of a Cash Transfer Programme for Schooling on Prevalence of HIV and Herpes Simplex Type 2 in Malawi: A Cluster Randomised Trial," *Lancet* 379: 1320–1329.

Bangsberg, D.R., Ware, N., and Simoni, J.M. (2006). "Adherence Without Access to Antiretroviral Therapy in Sub-Saharan Africa?," *AIDS* 20(1): 140–141.

Barberis, N.C. (2013). "Thirty Years of Prospect Theory in Economics: A Review and Assessment," *Journal of Economic Perspectives* 27(1): 173–96.

Catalani, C., Philbrick, W., Fraser, H., Mechael, P., and Israelski, D.M. (2013). "mHealth for HIV Treatment & Prevention: A Systematic Review of the Literature," *Open AIDS Journal* 7: 17–41.

Centers for Disease Control and Prevention. (2014). "Behavioral and Clinical Characteristics of Persons Receiving Medical Care for HIV Infection—Medical Monitoring Project," *HIV Surveillance Special Report 9.*

Centers for Disease Control and Prevention. (2015a). "HIV Among Gay and Bisexual Men," *HIV by Group.* Available at http://www.cdc.gov/hiv/group/msm/index.html (accessed January 16, 2017).

Centers for Disease Control and Prevention. (2015b). "HIV in the United States: At A Glance," *Statistics Center.* Available at http://www.cdc.gov/hiv/statistics/overview/ataglance.html (accessed January 16, 2017).

de Walque, D., Dow, W.H., Nathan, R., et al. (2012). "Incentivising Safe Sex: A Randomised Trial of Conditional Cash Transfers for HIV and Sexually Transmitted Infection Prevention in Rural Tanzania," *BMJ Open* 2:e000747. doi: 10.1136/bmjopen-2011-000747.

Foca, E., Odolini, S., Sulis, G., Calza, S., Pietra, V., Rodari, P., et al. (2014). "Clinical and Immunological Outcomes According to Adherence to First-Line HAART in an Urban and Rural Cohort of HIV-Infected Patients in Burkina Faso, West Africa," *BMC Infecious Diseases* 14.

Galárraga, O., Genberg, B.L., Martin, R.A., Barton Laws, M., and Wilson, I.B. (2013). "Conditional Economic Incentives to Improve HIV Treatment Adherence: Literature Review and Theoretical Considerations," *AIDS Behavior* 17(7): 2283–2292.

Garcia de Olalla, P., Knobel, H., Carmona, A., Guelar, A., Lopez-Colomes, J.L., and Cayla, J.A. (2002). "Impact of Adherence and Highly Active Antiretroviral Therapy on Survival in HIV-Infected Patients," *Journal of Acquired Immune Deficiency Syndrome* 30(1): 105–110.

Gardner, E.M., McLees, M.P., Steiner, J.F., Del Rio, C., and Burman, W.J. (2011). "The Spectrum of Engagement in HIV Care and Its Relevance to Test-and-Treat Strategies for Prevention of HIV Infection," *Clinical Infectious Diseases* 52(6): 793–800.

Genberg, B.L., Wilson, I.B., Bangsberg, D.R., Arnsten, J., Goggin, K., Remien, R.H., et al. (2012). "Patterns of Antiretroviral Therapy Adherence and Impact on HIV RNA Among Patients in North America," *AIDS* 26(11): 1415–1423.

Gray, W.N., Janicke, D.M., Fennell, E.B., Driscoll, D.C., and Lawrence, R.M. (2011). "Piloting Behavioral Family Systems Therapy to Improve Adherence Among Adolescents with HIV: A Case Series Intervention Study," *Journal of Health Psychology* 16(5): 828–842.

Halpern, S.D., Madison, K.M., and Volpp, K.G. (2009). "Patients as Mercenaries? The Ethics of Using Financial Incentives in the War on Unhealthy Behaviors," *Circulation-Cardiovascular Quality and Outcomes* 2(5): 514–516.

Hawkins, C. and Murphy, R. (2007). "Adherence to Antiretroviral Therapy in Resource-Limited Settings: Everything Matters," *AIDS* 21(8): 1041–1042.

HPTN. (2015). "HPTN 065: TLC-Plus: A Study to Evaluate the Feasibility of an Enhanced Test, Link to Care, Plus Treat Approach for HIV Prevention in the United States." Available at http://www.hptn.org/research_studies/hptn065.asp (accessed January 16, 2017).

IRINnews. (2012). UGANDA: Condom Use Infrequent Despite Rising HIV Rates, *IRINnews*. Available at http://www.irinnews.org/report/96359/uganda-condom-use-infrequent-despite-rising-hiv-rates (accessed January 16, 2017).

Jukes, M., Simmons, S., and Bundy, D. (2008). "Education and Vulnerability: The Role of Schools in Protecting Young Women and Girls From HIV in Southern Africa," *AIDS* 22: S41–S56.

Kahneman, D. (2011). *Thinking, Fast and Slow.* New York: Farrar, Straus and Giroux.

Kahneman, D. and Tversky, A. (1979). Prospect Theory: An Analysis of Decision under Risk. *Econometrica* 47(2): 263–291.

Kebba, A., Atwine, D., Mwebaze, R., Kityo, C., Nakityo, R., and Peter, M. (2002). "Therapeutic Responses to AZT + 3TC + EFV in Advanced Antiretroviral Naive HIV Type 1-Infected Ugandan Patients," *AIDS Research and Human Retroviruses* 18(16): 1181–1187.

Kobin, A.B. and Sheth, N.U. (2011). "Levels of Adherence Required for Virologic Suppression Among Newer Antiretroviral Medications," *Annals of Pharmacotherapy* 45(3): 372–379.

Kohler, H.P. and Thornton, R.L. (2011). "Conditional Cash Transfers and HIV/AIDS Prevention: Unconditionally Promising?" *The World Bank Economic Review* 26(2): 165–190.

Kongings, E., Yirga, A., Dilley, K., Gichangi, P., Arega, T., and Crandall, B. (2012). "Implications of Adopting New WHO Guidelines for Antiretroviral Therapy Initiation in Ethiopia," *Bulletin of the World Health Organization* 90(9): 659–663.

Li, J.Z., Paredes, R., Ribaudo, H.J., Svarovskaia, E.S., Kozal, M.J., Hullsiek, K.H., et al. (2012). "Relationship Between Minority Nonnucleoside Reverse Transcriptase Inhibitor Resistance Mutations, Adherence, and the Risk of Virologic Failure," *AIDS* 26(2): 185–192.

Linnemayr, S. (2015). "HIV Prevention Through the Lens of Behavioral Economics," *Journal of Acquired Immune Deficiency Syndrome* 68(4): e61–63.

Linnemayr, S. and Rice, T. (2016). "Insights from Behavioral Economics to Design More Effective Incentives for Improving Chronic Health Behaviors, with an Application to Adherence to Antiretrovirals," *Journal of Acquired Immune Deficiency Syndrome* 72(2): e50–52.

Linnemayr, S. and Stecher, C. (2015). "Behavioral Economics Matters for HIV Research: The Impact of Behavioral Biases on Adherence to Antiretrovirals (ARVs)," *AIDS Behavior* 19(11): 2069–2075.

Linnemayr, S., Stecher, C., and Mukasa, B. (2017). "Behavioral Economic Incentives to Improve Adherence to Antiretrovirals: Early Evidence from a Randomized Controlled Trial in Uganda," *AIDS*. Post Acceptance: January 4, 2017.

Loewenstein, G. (2005). "Hot-cold Empathy Gaps and Medical Decision Making," *Health Psychology* 24(4 Suppl): S49–56.

Lubinski, C. (2015). "IAS 2015: HPTN 068 and CAPRISA 007 Cash Incentive Studies Show Value of School, Financial Support, but also Show More Is Needed to Protect Young People from HIV." Available at http://sciencespeaksblog.org/2015/07/21/ias-2015-hptn-068-and-caprisa-007-cash-incentive-studies-show-value-of-school-financial-support-but-also-show-more-is-needed-to-protect-young-people-from-hiv/ (accessed January 16, 2017).

Manda, C. (2013). *Bang for Your Buck: STI Risk and Pregnancy Risk as Sources of the Price Premium for Unprotected Sex*. Munich. Available at http://econpapers.repec.org/paper/pramprapa/52864.htm (accessed January 16, 2017).

Mani, A., Mullainathan, S., Shafir, E., and Zhao, J. (2013). "Poverty Impedes Cognitive Function," *Science* 341(6149): 976–980.

Mermin, J., Lule, J., Ekwaru, J.P., Malamba, S., Downing, R., Ransom, R., et al. (2004). "Effect of Co-Trimoxazole Prophylaxis on Morbidity, Mortality, CD4-Cell Count, and Viral Load in HIV Infection in Rural Uganda," *The Lancet* 364(9443): 1428–1434.

Mermin, J., Were, W., Ekwaru, J.P., Moore, D., Downing, R., Behumbiize, P., et al. (2008). "Mortality in HIV-Infected Ugandan Adults Receiving Antiretroviral Treatment and Survival of their HIV-Uninfected Children: A Prospective Cohort Study," *The Lancet* 371(9614): 752–759.

Nachega, J.B., Mills, E.J., and Schechter, M. (2010). "Antiretroviral Therapy Adherence and Retention in Care in Middle-Income and Low-Income Countries: Current Status of Knowledge and Research Priorities," *Current Opinion in HIVand AIDS* 5(1): 70–77.

Nachega, J.B., Parienti, J.J., Uthman, O.A., Gross, R., Dowdy, D.W., Sax, P.E., et al. (2014). "Lower Pill Burden and Once-Daily Antiretroviral Treatment Regimens for HIV Infection: A Meta-Analysis of Randomized Controlled Trials," *Clinical Infectious Diseases* 58(9): 1297–1307.

Nakiyemba, A., Aurugai, D.A., Kwasa, R., and Oyabba, T. (2006). "Factors That Facilitate or Constrain Adherence to Antiretroviral Therapy Among Adults in Uganda: A Pre-Intervention Study," in A. Hardon, C. Hodgkin and R. Laing (eds), *From Access To Adherence: The Challenges of Antiretroviral Adherence*. Geneva: WHO, pp. 236–301.

Nyqvist, M., Corno, L., de Walque, D., and Svensson, J. (2015). *Using Lotteries to Incentivize Safer Sexual Behavior: Evidence from a Randomized Controlled Trial on HIV Prevention*. Washington DC: The World Bank, pp. 1–48.

Ortego, C., Huedo-Medina, T.B., Llorca, J., Sevilla, L., Santos, P., Rodriguez, E., et al. (2011). "Adherence to Highly Active Antiretroviral Therapy (HAART): A Meta-Analysis," *AIDS Behavior* 15(7): 1381–1396.

Parienti, J.J., Ragland, K., Lucht, F., de la Blanchardiere, A., Dargere, S., Yazdanpanah, Y., et al. (2010). "Average Adherence to Boosted Protease Inhibitor Therapy, Rather than the Pattern of Missed Doses, as a Predictor of HIV RNA Replication," *Clinical Infectious Diseases* 50(8): 1192–1197.

Pasternak, A.O., de Bruin, M., Jurriaans, S., Bakker, M., Berkhout, B., Prins, J.M., et al. (2012). "Modest Nonadherence to Antiretroviral Therapy Promotes Residual HIV-1 Replication in the Absence of Virological Rebound in Plasma," *Journal of Infectious Diseases* 206(9): 1443–1452.

Perneger, T.V. (1999). "Adjusting for Multiple Testing in Studies Is Less Important Than Other Concerns," *BMJ* 318(7193): 1288.

Pettifor, A., MacPhail, C., Nguyen, N., and Rosenberg, M. (2012). "Can Money Prevent the Spread of HIV? A Review of Cash Payments for HIV Prevention," *AIDS Behavior* 16(7): 1729–1738.

Sgaier, S.K., Reed, J.B., Thomas, A., and Njeuhmeli, E. (2014). "Achieving the HIV Prevention Impact of Voluntary Medical Male Circumcision: Lessons and Challenges for Managing Programs," *PLoS Med* 11(5).

Shuter, J. (2008). "Forgiveness of Non-Adherence to HIV-1 Antiretroviral Therapy," *Journal of Antimicrobial Chemotherapy* 61(4): 769–773.

Shuter, J., Sarlo, J.A., Stubbs, R.O., Rode, R.A., and Zingman, B.S. (2012). "Sequential Antiretroviral Adherence Measurement Using Electronic Bottle Cap Monitors in a Cohort of HIV-Infected Adults," *Journal of the International Association of Physicians of AIDS Care* 11(2): 94–97.

Simoni, J.M., Amico, K.R., Pearson, C.R., and Malow, R. (2008). "Strategies for Promoting Adherence to Antiretroviral Therapy: A Review of the Literature," *Current Infectious Diseases Reports* 10(6): 515–521.

The World Bank. (2002). "Education and HIV / AIDS—A Window of Hope," (p. 108).

Thirumurthy, H., Hayashi, K., Linnemayr, S., Vreeman, R.C., Levin, I.P., Bangsberg, D.R., et al. (2015). "Time Preferences Predict Mortality Among HIV-Infected Adults Receiving Antiretroviral Therapy in Kenya," *PLoS One* 10(12): e0145245.

Thornton, R.L. (2008). "The Demand for, and Impact of, Learning HIV Status," *American Economic Review* 98(5): 1829–1863.

Volpp, K.G., Troxel, A.B., Pauly, M.V., Glick, H.A., Puig, A., Asch, D.A., et al. (2009). "A Randomized, Controlled Trial of Financial Incentives for Smoking Cessation," *New England Journal of Medicine* 360(7): 699–709.

Witherspoon, E. (2015). "Behavioral Economics." Available at http://sph.unc.edu/cphm/unc-gillings-is-leading-mhealth-innovation/behavioral-economics/ (accessed January 16, 2017).

World Health Organization. (2015). "WHO | Guideline on when to Start Antiretroviral Therapy and on Pre-Exposure Prophylaxis for HIV," (p. 78).

World Health Organization and UNAIDS. (2007). *New Data on Male Circumcision and HIV Prevention: Policy and Programme Implications*. Geneva, Switzerland: WHO.

10

BEHAVIORAL ECONOMICS AND HEALTH BEHAVIORS AMONG THE POOR

Findings from developing country populations

Jill Luoto

Many, if not all, of the health behaviors covered in other chapters of this book that focus on populations in high-income countries also apply to populations in developing countries. After all, people everywhere make decisions on smoking, physical activity, eating, adhering to their medications, and screening for disease as well as interacting with insurers or providers. However, in developing countries, where markets often function poorly, institutions are weak, and poverty rates are high, many people must make even more critical decisions around health behaviors, often on a daily basis: whether to sleep under a mosquito net to prevent malaria, purchase and use a water treatment product to prevent waterborne diseases, cook with solid fuels and breathe the air pollutants or use an "improved" cookstove, etc. In developed countries these "decisions" have largely been taken out of the hands of individuals[1]—we turn on the faucet and expect treated drinking water, we turn on our electric or gas stoves that do not cloud our kitchens in smoke, and we have eradicated malaria from inside our borders. Public health measures such as improved sanitation and the provision of clean drinking water are two primary factors behind improvements in children's health in developed countries over the past 150 years (Currie, 2000). Individual or household behavior change was largely bypassed in achieving such environmental health gains.

Unfortunately, for poor populations in developing countries, there is both a larger overall disease burden—with environmental health risks still prevalent in addition to the non-communicable diseases (NCDs) such as cancers and cardiovascular disease that all populations face—and there are fewer resources to address those risks: health care systems are often low quality, and rates of insurance coverage are nearly always extremely low (Dupas, 2011; Das et al., 2008). One study found per capita health expenditures in high-income countries was

3000 USD per person in 2008, while in developing countries the corresponding figure was 30 USD per capita (Ke et al., 2011). Furthermore, education levels tend to be lower in the developing world. These factors contribute to the higher overall mortality risk in developing countries, particularly from parasitic and infectious diseases (World Health Organization (WHO), 2013). This puts a tremendous burden on these populations to manage multiple threats with scarce resources, and their health decision-making becomes all the more central to their overall welfare. Sadly, the vast majority of the disease burdens from both environmental health risks as well as from NCDs are preventable, if people behaved in a way that the traditional rational economic model would predict and their short-term behaviors better aligned with their long-term interests. Of course, the more decisions one must make, the more likely we are to see deviations from a rational model. And indeed, a review of health behavior in developing countries concludes, "when it comes to health behavior in developing countries, there are a substantial number of deviations from the neoclassical model" (Dupas, 2011, p. 455).

In this chapter we do not single out one health topic or behavior, but rather focus on the relatively large body of work in developing countries that examines behavioral economics applications or analyses on any kind of health topic or behavior. This includes studies both focused on behaviors of seeking or interacting with the health care system, as well as studies of more general health behaviors outside the formal health care system, whether preventive or curative in nature. We omit from this review studies of behavioral economics studies related to HIV/AIDS and risky sexual behaviors, as Chapter 9 in this volume focused specifically on HIV and contains many results from developing countries. However, even with this exclusion there are still a number of well-designed studies from developing countries focused on behavioral economics approaches to a variety of health problems, including attendance at recommended medical visits, adoption of insecticide-treated bednets or other preventive health products, and smoking cessation. The relatively strong emphasis on behavioral economics among studies from developing country settings is likely due to two factors. One, as already stated, there are simply more health decisions individuals must make in developing countries, arguably making behavioral economics approaches that much more applicable. Two, the recent expansive growth in the field of behavioral economics has co-occurred with recent expansive growth in the field of development economics and its affinity for randomized field experiments and trials, where it is often comparatively cost-effective to test a given behavior or problem. In total this has resulted in a rapidly growing field of behavioral economics findings on health applications from the developing world (Datta and Mullainathan, 2014; Ashton et al., 2015).

Yet no matter where the study or what the health problem is, the majority of approaches and policies designed to address that problem continue to take on a traditional economic flavor such as pricing and informational interventions. Though the potential for behavioral economics may be significant, the evidence base for its promise is still predominantly made up of small-scale randomized controlled trials—policy interventions or scaled approaches based on behavioral economics principles are few and far between in the developing world, as anywhere else. Some critics argue that the evidence base for "nudging" as a policy tool to improve population health is far behind popular enthusiasm for the approach (Marteau et al., 2011). Others point out that the effect sizes from behavioral economics or "nudge" policies are a pittance when compared to the scale of many problems societies face (Porter, 2016). One review of public health strategies to combat unhealthy diets, physical inactivity, and obesity in low- and middle-income countries argued that more traditional approaches such as informational and pricing interventions and regulation of marketing to children would be most likely to produce the largest impacts in the shortest amount of time (Cecchini et al., 2010). (Note, however, that this review did not compare such approaches to behavioral economics policies or nudges.) In any event, these criticisms put behavioral economics as a field in a tough spot when it comes to the promise it holds for promoting health in developing countries. On the one hand, in the face of market failures, widespread poverty, oftentimes corrupt public sectors, and the sheer number of health decisions left in the hands of individuals, any interventions or policies that are cheap (or free) to implement and that can overcome such obstacles to improve outcomes—even if only by a small amount—seem worthwhile. On the other hand, even in the best case scenarios, nudge approaches built on insights from behavioral economics are often just that, nudges; arguably, they alone are not likely to overcome the sheer scale of many of the health problems that societies face, and particularly in the developing world. Large-scale public health investments may ultimately be more cost-effective (Dupas, 2011).

The rest of this chapter presents a brief overview of the traditional versus behavioral economics approaches to studying the determinants of health behaviors in developing countries. It then presents a synopsis of the evidence base for behavioral economics studies to improve health decision-making that takes place in a developing country setting. It concludes by identifying some areas for future research that could prove valuable in advancing the field.

Traditional approaches to health promotion in developing countries

The neoclassical rational model built upon utility maximization and expected utility theory generally assumes that people have full information and know what is best for themselves, and make decisions that reflect their true preferences subject to their available resources (time, information, and prices of goods as well as labor). In instances where people err in their decision-making, the rational model generally assumes that they will learn from their mistakes—that is, any information that was once lacking can be gained through experience. In the context of health decisions, which often involve multiple time periods, the standard model assumes that people are time-consistent and the marginal rate of substitution for consumption between any two time periods is constant over time. That is, people have the same preferences regardless of the timing of the decision, and will exponentially discount future periods at some (constant) rate relative to the present period.

Traditional policy tools informed by the neoclassical model to address health behaviors thus generally consist of interventions to correct any asymmetric information in the marketplace, incentivizing the intended behavior directly, such as through financial incentives; pricing interventions, such as through taxes or subsidies; and mandates or other forms of legislation.

Informational interventions

In the context of informational interventions to correct a problem of incomplete information that negatively affects health, Madajewicz et al. (2007), Jalan and Somanathan (2008), and Bennear et al. (2011) find evidence that providing information about water quality can positively impact households' water-related treatment or collection behaviors, and Cohen (2014) similarly finds a role for information affecting households' investments in anti-malarial drugs. In sum, these studies suggest that incomplete information can be one barrier to the uptake of preventive health behaviors—populations may not know about the dangers of a certain behavior, or may not know ways to overcome those dangers. However, Kremer and Miguel (2007) did not find any effects from an informational intervention aimed at improving intestinal worm infections in Kenya, and Meredith et al. (2013) also find no effect from sharing information on investment in preventive health products across four randomized field trials focused on different health products. The effectiveness of informational interventions therefore seems to depend on both the context, as well as on how that information is provided (Dupas, 2011)—vivid and salient messages communicated in a clear and simple fashion tend to be most effective, and sometimes their effects are not clearly due to the information communicated so much as the reminder or salience of that information (Rosenberg, 2016). In that case, behavioral insights may be able to help inform the best design for informational interventions to maximize their impact (Datta and Mullainathan, 2014).

Incentives for healthy behaviors

Perhaps the best-known and most widely implemented real-world examples of the use of financial incentives to reward healthy behavior come from conditional cash transfer (CCT) programs, which pay people (usually the poor) upon completion of a stated goal or behavior. Most often, the conditions for such transfers surround school attendance or uptake of preventive health services such as check-ups or child health immunizations. CCTs programs have grown tremendously throughout many developing countries, including several in Latin America, and even some developed countries. Their focus has expanded over time as well in terms of the behaviors upon which the conditions are designed, and by now CCT programs have successfully increased the use of contraception, delayed marriage, and increased the use of antenatal care (Ashraf et al., 2014; Baird et al., 2010). Overall, the evidence suggests that such CCT programs can be effective in increasing the uptake of health services, as well as the promotion of health behaviors (Lagarde and Palmer, 2007; Ranganathan and Lagarde, 2012).

However, critics of CCT programs argue that they can be expensive, and risk creating perverse incentives as well as undermining people's intrinsic motivation to perform the rewarded behavior. Indeed, reviews of CCT programs have found little evidence that long-term change in behavior will be sustained once the incentives are withdrawn (Ranganathan and Lagarde, 2012).

Pricing interventions

As the vast majority of those living in developing countries are poor, pricing interventions such as taxes or subsidies are one of the most common interventions used to encourage uptake of health behaviors or products by overcoming liquidity constraints. For preventive health products, there is by now a wealth of evidence from randomized trials in developing countries that price is a significant determinant of take-up, with households very sensitive to even small changes in price (Holla and Kremer, 2009). For instance, Meredith et al. (2013) find, across four experiments, that demand for health products drops off rapidly when prices go from zero to any positive amount. Similar findings can be found for a water treatment product (Ashraf et al., 2010) and anti-malarial mosquito nets (Cohen and Dupas, 2010; Dupas, 2009, 2014). Okeke et al. (2013) similarly find that demand for a cancer screening test is very sensitive to the price of any subsequent necessary treatment, in the event of a positive diagnosis.

There is also some early evidence from large-scale policy changes that well-designed pricing interventions can promote health behaviors. Mexico introduced an excise tax on sugar sweetened beverages to combat high rates of consumption coupled with rapidly rising rates in obesity and related diseases (Barquera et al., 2008). Early reviews suggest it is having positive effects in reducing soda purchases by 12 percent (Colchero et al., 2016), though its impact on health remains unclear at this point. A large-scale private insurer in South Africa that offered its members cash-back rebates on healthy food purchases also found that participation in the program led to increases in purchases of healthy foods and decreases in purchases of less-desirable foods (Sturm et al., 2013).

Though pricing interventions have shown promise, even when health products or health care access are given away for free, take-up is still not 100 percent. For instance, even after making supply-side improvements to ensure health care clinics in India are reliably open and staffed as intended, and even when immunizations are free, more than 80 percent of households do not fully immunize their children (Banerjee et al., 2010; Dupas, 2011). As another example, households in Kenya and Bangladesh that were provided a variety of free safe water products, along with repeated informational appeals for reasons to use them, still did not achieve anywhere near 100 percent adoption with any single product (Luoto et al., 2011; Albert et al., 2010). These findings suggest that price is one of many barriers to take-up of preventive health products and behaviors, potentially leaving room for behavioral economics insights to improve on such outcomes.

Mandates or other legislation

Though mandates are certainly one way to enforce compliance with a given health behavior, such mandates must be enforced in order to be effective, and must have the expected outcome. In developing countries, where institutions are often weak, resources for enforcement may be scarce, and quality of supply may be inadequate, mandates may have questionable value in actually improving health outcomes. For example, Godlonton and Okeke (2016) evaluate a ban on traditional birth attendants that was introduced in Malawi in 2007 and aimed to reduce the use of informal health care services as one way to improve maternal and child health outcomes. They find significant numbers of mothers shifting from informal to formal care, but neonatal mortality improves only among those women who live near a high-quality health care facility, with no improvement on average. They attribute this result to the poor average quality of clinics in Malawi. Ranganathan and Lagarde (2012) in their review of 13 CCT programs from various countries also argue that increasing demand (via whatever means), without a concurrent improvement in the quality of supply, risks rewarding the health behaviors but ultimately not improving health (Dupas, 2011). This suggests that mandates to encourage a certain behavior on the demand-side must be carefully considered to ensure the supply-side has the ability to adequately meet that demand to improve health. (We do not discuss the effectiveness of supply-side interventions to improve the delivery of health care or health products, as this is outside the scope of this chapter.)

Why behavioral economics?

Though there are examples of success from traditional efforts to encourage health behavior change, the general consensus in many settings around the world is that they have ultimately fallen short of the scale of health behavior changes needed (WHO, 2008). The limited effectiveness of traditional tools may be due to shortcomings of the rational model that largely informs their design, which cannot explain some commonly observed health phenomena such as the stark rise in obesity rates worldwide, or the fact that some people continue to drink excessively or smoke even when they wish to quit. Or, more relevant to very poor settings, the fact that many households intend to vaccinate their children, or treat their drinking water, yet fail to follow through on those plans.

These shortcomings have heightened the search for effective interventions to encourage health behavior change, as well as a search outside of traditional economics for tools to achieve such change.

The field of behavioral economics has gained much attention in recent years as an exciting advance in our ability to explain people's choices in many contexts where the rational model can sometimes fall short. Whereas the rational model is a normative model of decision-making, behavioral economics comprises a descriptive theory of decision-making that describes how people actually make decisions versus how they should. Importantly, behavioral economics is not a substitute for the traditional model, but rather a complement or an extension of neoclassical economics that incorporates insights from psychology and other social sciences to improve our ability to understand people's decision-making tendencies in certain scenarios.

In light of the findings that people regularly and systematically deviate from a rational model of behavior, and that such deviations can hurt their own long-term interests, much of the recent work in behavioral economics has focused on developing interventions that expressly take such decision-making tendencies into account. The idea is that by knowing these decision-making heuristics or 'rules of thumb' and rearranging the context of the problem with such patterns in mind, these same errors that usually hurt people can potentially be harnessed to help them instead. That is, behavioral economists have developed a variety of techniques (sometimes called "nudges") to encourage individuals to make better choices that are informed by behavioral economics insights into how people actually make decisions.

In this section, we present some of the evidence on the most common and successful behavioral economics interventions, many based on nudges, to address various health behaviors and that take place in developing country settings. Though most of the evidence cited was expressly designed with insights from behavioral economics in mind, we also cite other studies of health behaviors in developing countries where there is a plausible explanation from behavioral economics that could explain their findings.

Using behavioral economics to promote health in developing country settings

To date, much of the focus of behavioral economics studies from developing countries is on financial decision-making such as testing ways to encourage savings (Ashraf et al., 2006; Dupas and Robinson, 2013). Though ostensibly a financial, not health, behavior, many of these studies focus on savings specifically earmarked for health and investments in health products such as mosquito nets or water treatment products. This is because a significant unsolved puzzle in health and development is why we commonly observe high levels of curative health expenditures matched with low levels of preventive expenditures among the poor (Dupas, 2011).

Commitment savings accounts to address present bias

One commonly evoked explanation for this pattern is present bias—the tendency to over-weight the present period in decision-making at the expense of the long-run consequences. That is, the poor intend to save or to commit a certain behavior, but temptations today may prevent them from doing so. One of the most common behavioral economics tools designed to overcome present bias is a commitment savings account, which is designed to appeal to people's self-awareness of their problems with self-control. "Sophisticated" present biased types who understand this tendency will commit themselves today to a savings behavior that "ties their hands" and doesn't allow their future selves to break that commitment. There is by now a large body of evidence that such commitment contracts can positively encourage a variety of health behaviors, including smoking cessation in the Philippines (Gine et al., 2010), general health savings accounts in Kenya (Dupas and Robinson, 2013), uptake of long-lasting insecticide-treated bednets in India (Tarozzi et al., 2010), and attendance at regular check-ups among hypertensives in India (Bai et al., 2015).

The Tarozzi et al. (2010) study is a good example of a creative commitment savings product that was found to improve health behaviors among the poor. Clients of a microfinance institution (MFI) in Orissa, India, were allowed to purchase insecticide-treated bednets on credit, but they were given the choice of committing themselves in advance to purchase the nets alone, or to purchase the nets with two prepaid insecticide retreatments 6 and 12 months later. The latter sales contract formed a commitment contract for long-lasting insecticide treatment and enjoyed high demand. Of those purchasing the nets, 53 percent opted for the commitment contract to agree in advance for the retreatment. When the researchers returned to the villages to retreat the nets 6 months later, the households who had not taken up the original offer were allowed to pay for retreatment on the spot, but just 30 percent did so.

Micro-incentives to address present bias

Another tool to overcome present bias can be small ("micro") incentives—giving people a small reward today to commit a behavior they may otherwise procrastinate doing. Though financial incentives such as CCT programs are classified as tools informed by traditional economics above, the incentives they offer may address present bias that prevents such health behaviors from occurring in the first place. More generally, since many health promotion behaviors are characterized by effort today with benefits delayed (and uncertain), such micro-incentives can address present bias without significantly altering one's budget constraint, thus qualifying as more of a behavioral, than a traditional, policy tool. In some cases those incentives do not even need to be financial. For example, behavioral economics recognizes the importance of social norms in our decision-making (Cialdini, 1993; Thaler and Sunstein, 2009), and micro-incentives can capitalize on this insight. Ashraf, Bandiera, and Jack (2012) find that non-cash incentives in the form of peer recognition were more effective than financial incentives to motivate health extension workers in Zambia to promote and sell female condoms. As another example of non-financial micro-incentives, Banerjee et al. (2010) consider the problem that, despite the benefits of immunization to improve child survival, and despite immunizations being free in public health facilities, the majority of children do not receive the basic package of immunizations. They hypothesize different reasons for this underutilization including procrastination by parents (usually mothers), who may delay bringing their children in for an immunization or may drop out before their child has received the full course of immunizations. Banerjee et al. (2010) test this hypothesis by providing mothers in some randomly selected villages with a small incentive in the form of 1 kilogram of lentils per immunization, and a set of metal meal plates upon completion of the full immunization course. The non-financial lentil incentives significantly increase rates of immunization from 18 percent among villages that received reliable supply-side improvements, to 39 percent among villages that additionally receive the demand-side incentives (Banerjee et al., 2010).

The role of limited attention in health decision-making

Though present bias is the most common behavioral explanation for why poor households underinvest in health products or savings, some suggest that limited attention may also play a role. That is, while the traditional economic model assumes people have unlimited cognitive capacities, a number of studies from psychology suggest that attention is a limited resource. When attention is limited, the salience, or vividness of a signal will determine what people's limited cognitive capacities attend to. Recent work suggests that poverty itself can create additional psychological burdens, which can reduce the available bandwidth in decision-making among the poor (Shah et al., 2012; Mani et al., 2013). Thus, the poor may not invest in preventive health products because their attention is focused on more immediate needs and concerns, and they do not have the cognitive bandwidth to consider the costs and benefits of a purchase decision (Datta and Mullainathan, 2014; Spears, 2011). Spears (2011) tests this idea for the purchase of soap among poor households in India. A randomly selected treatment group was asked to think about the value of money and then was offered soap at either 3 or 15 rupees. The control group was simply offered soap at either 3 or 15 rupees. The treatment group had a somewhat lower price elasticity of demand for the soap (-0.5 versus -0.7), suggesting decision costs may play a role in dampening demand (Kremer and Glennerster, 2012).

Behavioral economics can encourage greater effort to improve health ...

Of course, not all health decisions involve purchase decisions. Many health behaviors are in fact free, but all have some kind of time or effort cost. For instance, even after purchasing a safe water product one must use it (repeatedly) in order to derive any benefits. An increasing number of behavioral economics studies have looked at these "free" behaviors among the poor in the developing world to test new insights or ways to promote health behavior change.

In the context of safe water behaviors, Luoto et al. (2014) provide free safe water products to households in Bangladesh and Kenya and deliver different marketing pitches to test different persuasion tactics informed by behavioral economics. In both settings they test both a "framing" treatment that appeals to loss aversion (as compared to a positively framed message only emphasizing the gains from treatment), as well as a treatment that asked households to make a verbal pledge to treat one's drinking water, meant to appeal to the finding that people often work hard to remain consistent with a prior public commitment (Cialdini, 1993; Greenwald et al., 1987). They find in both Kenya and Bangladesh samples that both "nudges" incrementally increase rates of usage of the free products, and appear addictive in nature, which drives up rates further in tandem.

... Or, behavioral economics can reduce the effort needed to improve health

Another approach tested to deliver safe water informed by behavioral economics principles is creating a "smarter" default to bypass the need for individuals to remember and follow through on daily water treatment, capitalizing on the status quo bias, or the behavioral economics finding that people have a disproportionate preference for the default option in a menu of choices. Ahuja et al. (2010) installed automated chlorine dispensers at water collection points throughout western Kenya. These dispensers attempt to make water treatment as automatic as possible and remove most of the steps that individuals must undertake in order to treat one's water, by dispensing correctly measured amounts of chlorine, for free, at the press of a button, and locating them where people already go to collect water. When paired with a community promoter to explain the dispensers, take-up of chlorine increased by 53 percentage points compared to a control group (Ahuja et al., 2010).

Devoto et al. (2012) also find large effects from offering a smarter design for the default in choosing whether to connect to the municipal piped water among households in Morocco. Nearly 70 percent of those helped with the administrative process agreed to sign up for a piped water connection, compared to just 10 percent in a control group. Though not a direct health study, it suggests that even small administrative hassles can have large effects on take-up rates for various programs that ultimately can affect one's health (such as drinking water).

Small inconveniences can have big consequences

Further evidence on the power of defaults comes from India, where Banerjee and Duflo (2011) had designed a program for fortifying flour with iron to improve nutritional outcomes for children. The planned design was that households would only have to tell millers once whether they wanted enriched flour, and the miller would repeat that action each time they returned. Instead, the millers did the opposite, and forced households to specify each time whether they wanted iron-fortified flour. This change in the default resulted in significant drop-out from the program (Datta and Mullainathan, 2014). Yet more potential evidence on the power of poorly designed defaults to have large negative effects on behavior is the finding that women across 60 developing countries are, surprisingly, likely to report they have discontinued use of contraceptives because of method inconvenience (Ashton et al., 2015; Ali et al., 2012).

Behavioral economics insights can appear even when not planned

Many health promotion studies from developing countries, in effect, incorporate behavioral design principles, even if they are not formally labeled as behavioral economics studies. One example comes from a quasi-experimental evaluation of a program called Berhane Hewan, which aimed to delay marriage among adolescent girls and empower them to continue their education in Ethiopia. Early marriage by adolescent girls can lead to their having less negotiating status in the home as girls are much younger than their spouses, as well as early and risky first births, and worse outcomes for the infants born to very young mothers (Raj, 2010; Raj et al., 2009). Focused on 10–14-year-olds, Berhane Hewan included a public commitment by parents and their daughters to delay marriage (capitalizing again on the commitment–consistency principle) and offered a goat upon successful completion of the program to help offset costs of delaying marriage (Ashton et al., 2015). Those girls aged 10–14 years, exposed to the program, were three times more likely to be in school at the endline survey two years later than a control group, and were one-tenth as likely to have ever been married (Erulkar and Muthengi, 2007).

Behavioral economics is not a panacea

Looking across the spectrum of health behaviors studied and the various locales, it is clear that the growing field of behavioral economics has become a promising source for informing the design of interventions and policies intended to improve uptake of health behaviors among developing country populations. Because many behavioral economics tools or policies simply realign the choice architecture of a given problem to maximize welfare, they can often be low-cost to implement and therefore of particular value in low-resource developing country settings. However, the excitement for a new field and source for new ideas does not mean that behavioral economics will solve all of our behavioral shortcomings that can seem endemic to human nature everywhere. Moreover, the enthusiasm for behavioral economics should be tempered against the realization that its evidence base is still just beginning to grow, and still predominantly relies on pilot studies and small-scale randomized trials (and most of which still are located in high-income countries) (Marteau et al., 2011).

There is reason to question the ability of behavioral economics to overturn larger problems on a greater scale, as nudges are not meant to move mountains—they are meant only to push a few people over the cliff. Eldar Shafir, a professor of psychology at Princeton and a prominent leader in the field of behavioral economics, was quoted in an interview with the New York Times as saying, "The insights from behavioral economics are beautiful from a research perspective. But its popularity no doubt comes from a combination of lack of funds and political helplessness." (Porter, 2016). If lack of funds and political helplessness are particularly characteristic of many developing countries, this could explain the enthusiasm for the approach in such settings. However, these settings are often precisely where more than just a nudge is needed, and this should be taken into account in any intervention design or program.

Conclusion

The high disease burden in developing countries coupled with weak institutions make insights from behavioral economics hold particular value as a creative means to encourage health promotion and potentially to save many lives. Finding ways to reduce the sheer prevalence of risky behaviors underlying much of these disease burdens—ranging from tobacco use and underuse of preventive care to failure to treat one's water or invest in a mosquito net—has the potential to reduce much of the dual disease burdens that afflict poor countries (WHO, 2013; Dupas, 2011). Traditional policy tools informed by the rational model are increasingly viewed as not sufficient at modifying such behaviors on their own (Europe, 2008). The growing field of behavioral economics has become a promising additional source for informing the design of policies intended to increase the uptake of healthy behaviors (Datta and Mullainathan, 2014).

The potential for behavioral economics as a source for helping to solve such health problems is thought by many to be large and promising and to primarily lie in the field's recognition of the way the social and physical environments shape our behavior. However, to date the field is still emerging, and the evidence base for its potential to address health problems in the developing world is mostly made up of small-scale randomized field studies. Its ability to change behaviors at the level of general populations or to achieve long-term sustained behavior change remains largely unanswered (Marteau et al., 2011). Looking forward, Datta and Mullainathan (2014) argue that behavioral economists working in development must move away from "boutique" pilot studies that isolate singular behavioral channels and instead encourage using insights that can be applied and have impact at scale.

In summary, behavioral economics insights can improve the designs of many policies and programs that aim to improve the health of poor populations in developing countries; any policies or programs that ignore those insights on how people make decisions, do so at their own peril. However, behavioral economics insights, alone, are not a panacea. The more decisions we can take out of the hands of individuals in developing countries, such as through large-scale public health investments like sanitation systems and delivery of piped water, the more likely we are to help poor populations better manage those health decisions that remain under their control. Behavioral economics can help improve demand-side health behaviors, but should not alone be the final solution.

Note

1 An exception being cases like lead-contaminated drinking water in Flint, Michigan, publicized in 2016. Some material in this chapter appears in the author's Technical Note for the Inter-American Development Bank publication,"Behavioral Economic Guidelines with Application for Health Interventions." Available at https://publications.iadb.org/bitstream/handle/11319/6503/Behavioral_economics_guidelines_with_applications_for_health_interventions.pdf?sequence=1 (accessed January 3, 2017).

References

Ahuja, A., Kremer, M., and Zwane, A. (2010). "Providing Safe Water: Evidence from Randomized Evaluations," *Annual Review of Resource Economics* 2: 237–256.

Albert, J., Luoto, J., and Levine, D.I. (2010). "End-User Preferences for and Performance of Competing POU Water Treatment Technologies Among the Rural Poor of Kenya," *Environment, Science and Technology* 44(12): 4426–4432. Available at http://www.ncbi.nlm.nih.gov/pubmed/20446726 (accessed January 3, 2017).

Ali, M.M., Cleland, J., and Shah, I.H. (2012). "Causes and Consequences of Contraceptive Discontinuation: Evidence from 60 Demographic and Health Surveys," *World Health Organization.*

Ashraf, N., Bandiera, O., and Jack, B.K. (2012). *No Margin, No Mission? A Field Experiment on Incentives for Pro-Social Tasks.* Available at http://papers.ssrn.com/sol3/papers.cfm?abstract_id=2013825 (accessed January 3, 2017).

Ashraf, N., Berry, J., and Shapiro, J. (2010). "Can Higher Prices Stimulate Product Use? Evidence from a Field Experiment in Zambia," *American Economic Review* 100(5): 2383–2413.

Ashraf, N., Field, E., and Lee, J. (2014). "Summary—Household Bargaining and Excess Fertility: An Experimental Study in Zambia," *American Economic Review* 104(7, July): 1–4.

Ashraf, N., Karlan, D., and Yin, W. (2006). "Tying Odysseus to the Mast: Evidence from a Commitment Savings Product in the Philippines," *The Quarterly Journal of Economics* 121(2): 635–672.

Ashton, L., Giridhar, N., Holcombe, S.J., Madon, T., and Turner, E. (2015). "A Review of Behavioral Economics in Reproductive Health," *Version January,* 20. Available at http://www.beri-research.org/wp-content/uploads/2015/01/BERI-White-Paper_version_1.20.15.pdf (accessed January 3, 2017).

Bai, L., Handel, B., Miguel, E., and Rao, G. (2015). "Self-Control and Chronic Illness: Evidence from a Field Experiment," Working Paper. Available at https://dl.dropboxusercontent.com/u/10518610/EHP_slides.pdf (accessed January 3, 2017).

Baird, S., Chirwa, E., McIntosh, C., and Ozler, B. (2010). "The Short-term Impacts of a Schooling Conditional Cash Transfer Program on the Sexual Behavior of Young

Women," *Health Economics* 19(S1): 55–68. Available at http://dx.doi.org/10.1002/hec.1569 (accessed January 3, 2017).

Banerjee, A. and Duflo, E. (2011). *Poor Economics: A Radical Rethinking of the Way to Fight Global Poverty*. New York: Public Affairs.

Banerjee, A., Duflo, E., Glennerster, R., and Kothari, D. (2010). "Improving Immunisation Coverage in Rural India: Clustered Randomised Controlled Evaluation of Immunisation Campaigns With and Without Incentives," *British Medical Journal* 340(May17, 1): c2220–c2220. Available at http://www.bmj.com/cgi/doi/10.1136/bmj.c2220 (accessed January 3, 2017).

Barquera, S., Hernandez-Barrera, L., Tolentino, M.L., Espinosa, J., Ng, S.W., Rivera, J.A., Popkin, B.M., et al. (2008). "Energy Intake from Beverages Is Increasing Among Mexican Adolescents and Adults," *The Journal of Nutrition* 138 (12): 2454–2461. Available at http://jn.nutrition.org/content/138/12/2454.abstract (accessed January 3, 2017).

Bennear, L., Tarozzi, A., Pfaff, A.A., Soumya, H.B., Ahmed, K.M., and van Geen, A. (2011). "Bright Lines, Risk Beliefs, and Risk Avoidance: Evidence from a Randomized Intervention in Bangladesh," Working Paper. Available at https://papers.ssrn.com/sol3/papers.cfm?abstract_id=1695115 (accessed January 3, 2017).

Cecchini, M., Sassi, F., Lauer, J.A., Lee, Y.Y. Guajardo-Barron, V., and Chisholm D. (2010). "Tackling of Unhealthy Diets, Physical Inactivity, and Obesity: Health Effects and Cost-Effectiveness," *The Lancet* 376(9754): 1775–1784. Available at http://www.sciencedirect.com/science/article/pii/S0140673610615140 (accessed January 3, 2017).

Cialdini, R.B. (1993). *Influence: The Psychology of Persuasion*. New York: Morrow.

Cohen, D. (2014). *A Big Fat Crisis—The Hidden Forces Behind the Obesity Epidemic—And How We Can End It*, 1st edn. New York: Nation Books.

Cohen, J. and Dupas, P. (2010). "Free Distribution or Cost Sharing? Evidence from a Randomized Malaria Prevention Experiment," *Quarterly Journal of Economics* CXXV(1).

Colchero, M.A., Popkin, B.M., Rivera, J.A., and Ng, S.W. (2016). "Beverage Purchases from Stores in Mexico Under the Excise Tax on Sugar Sweetened Beverages: Observational Study," *BMJ*, 352. Available at http://www.bmj.com/content/352/bmj.h6704.abstract (accessed January 3, 2017).

Currie, J. (2000). "Chapter 19 Child Health in Developed Countries," *Handbook of Health Economics* 1(PART B): 1053–1090.

Das, J., Hammer, J., and Leonard, K. (2008). "The Quality of Medical Advice in Low-income Countries," *Journal of Economic Perspectives* 22(2): 93–114.

Datta, S. and Mullainathan, S. (2014). "Behavioral Design—A New Approach to Development Policy," *Review of Income and Wealth*, 60(1). Available at http://www.cgdev.org/files/1426679_file_Datta_Mullainathan_Behavioral_Design.pdf (accessed January 3, 2017).

Devoto, F., Duflo, E., Dupas, P., Pariente, W., and Pons, V. (2012). "Happiness on Tap: Piped Water Adoption in Urban Morocco," *AEJ: Economic Policy* 4(4): 68–99. Available at http://www.nber.org/papers/w16933 (accessed September 1, 2011).

Dupas, P. (2009). "What Matters (and What Does Not) in Households' Decision to Invest in Malaria Prevention?" *American Economic Review: Papers and Proceedings* 99(2): 224–230.

Dupas, P. (2011). "Health Behavior in Developing Countries," *Annual Review of Economics* 3: 425–449.

Dupas, P. (2014). "Short-Run Subsidies and Long-Run Adoption of New Health Products: Evidence from a Field Experiment," *Econometrica* 82(1): 197–228. Available at http://www.stanford.edu/~pdupas/Subsidies&Adoption.pdf (accessed January 3, 2017).

Dupas, P. and Robinson, J. (2013). "Why Don't the Poor Save More? Evidence from Health Savings Experiments," *American Economic Review* 103(4): 1138–1171.

Erulkar, A.S. and Muthengi, E. (2007). "Evaluation of Berhane Hewan—A Pilot Program to Promote Education and Delay Marriage in Rural Ethiopia," *International Perspectives on Sexual and Reproductive Health* 35(1): 6–14. Available at http://www.popcouncil.org/pdfs/Ethiopia_EvalBerhaneHewan.pdf (accessed January 3, 2017).

Europe, W.R.C. (2008). *Behaviour Change Strategies and Health: The Role of Health Systems.*

Gine X., Karlan, D., and Zinman, J. (2010). "Put Your Money Where Your Butt Is: A Commitment Contract for Smoking Cessation," *World Bank Economic Review* 22(3): 539–566.

Godlonton, S. and Okeke, E.N. (2016). "Does a Ban on Informal Health Providers Save Lives? Evidence from Malawi," *Journal of Development Economics* 118: 112–132. Available at http://www.sciencedirect.com/science/article/pii/S0304387815001029 (accessed January 3, 2017).

Greenwald A.G., Carnot, C., Beach, R., and Young, B. (1987). "Increasing Voting Behavior by Asking People if They Expect to Vote," *Journal of Applied Psychology* 72: 315–318.

Holla, A. and Kremer, M. (2009). "Pricing and Access: Lessons from Randomized Evaluations in Education and Health," *Center for Global Development*, Working Paper, (158).

Jalan, J. and Somanathan, E. (2008). "The Importance of Being Informed: Experimental Evidence on Demand for Environmental Quality," *Journal of Development Economics* 87: 14–28.

Ke, X., Saksena, P., and Holly, A. (2011). "The Determinants of Health Expenditure: A Country-Level Panel Data Analysis," p. 26. Available at http://www.who.int/health_financing/documents/report_en_11_deter-he.pdf (accessed January 3, 2017).

Kremer, M. and Glennerster, R. (2012). "Improving Health in Developing Countries: Evidence from Randomized Evaluations," in *Handbook of Health Economics, Volume 2*. Netherlands: Elsevier B.V., pp. 201–315. Available at http://dx.doi.org/10.1016/B978-0-444-53592-4.00004-9 (accessed January 3, 2017).

Kremer, M. and Miguel, E. (2007). "The Illusion of Sustainability," *The Quarterly Journal of Economics* 122(3): 1007–1065.

Lagarde, M., Haines, A., and Palmer, N. (2007). "Conditional Cash Transfers for Improving Uptake of Health Interventions in Low- and Middle-income Countries: A Systematic Review," *JAMA: The Journal of the American Medical Association* 298(16): 1900–1910. Available at http://dx.doi.org/10.1001/jama.298.16.1900 (accessed January 3, 2017).

Luoto, J., Levine, D., Albert, J., and Luby, S. (2014). "Nudging to Use: Achieving Safe Water Behaviors in Kenya and Bangladesh," *Journal of Development Economics* 110: 13–21.

Luoto, J., Najnin, N., Mahmud, M., Albert, J., Islam, M.S., Luby, S., Unicomb, L., and Levine, D.I. (2011). "What Point-of-Use Water Treatment Products Do Consumers Use? Evidence from a Randomized Controlled Trial among the Urban Poor in Bangladesh," *PLoS ONE*, 6(10). Available at http://www.plosone.org/article/info%3Adoi%2F10.1371%2Fjournal.pone.0026132 (accessed January 3, 2017).

Madajewicz, M., Pfaff, A., van Geen, A., Graziano, J., Hussein, I., Momotaj, H., Sylvi, R., and Ahsan, H. (2007). "Can Information Alone Change Behavior? Response to Arsenic Contamination of Groundwater in Bangladesh," *Journal of Development Economics* 84: 731–754.

Mani, A., Mullainathan, S., Shafir, E., and Zhao, J. (2013). "Poverty Impedes Cognitive Function," *Science* 341(6149): 976–980. Available at http://www.sciencemag.org/content/341/6149/976.abstract (accessed January 3, 2017).

Marteau, T.M., Ogilvie, D., Roland, M., Suhrcke, M., and Kelly, M.P. (2011). "Judging Nudging: Can Nudging Improve Population Health?" *BMJ*, 342.

Meredith, J., Robinson, J., Walker, S., and Wydick, B. (2013). "Keeping the Doctor Away: Experimental Evidence on Investment in Preventative Health Products," *Journal of Development Economics* 105: 196–210.

Okeke, E., Adepiti, A., and Ajenifuja, K. (2013). "What Is the Price of Prevention? New Evidence from a Field Experiment," *Journal of Health Economics* 32(1): 207–218.

Porter, E. (2016). "Nudges Aren't Enough for Problems Like Retirement Savings," *New York Times*. Available at http://www.nytimes.com/2016/02/24/business/economy/nudges-arent-enough-to-solve-societys-problems.html?login=email&hpw&rref=business&action=click&pgtype=Homepage&module=well-region&region=bottom-well&WT.nav=bottom-well&_r=1&mtrref=undefined (accessed January 3, 2017).

Raj, A., Saggurti, N., Balaiah, D., and Silverman, J.G. (2009). "Prevalence of Child Marriage and Its Effect on Fertility and Fertility-control Outcomes of Young Women in India: A Cross-sectional, Observational Study," *The Lancet* 373(9678): 1883–1889. Available at http://www.sciencedirect.com/science/article/pii/S0140673609602464 (accessed January 3, 2017).

Raj, A. (2010). "When the Mother Is a Child: The Impact of Child Marriage on the Health and Human Rights of Girls," *Archives of Disease in Childhood*. Available at http://adc.bmj.com/content/early/2010/10/07/adc.2009.178707.short (accessed January 3, 2017).

Ranganathan, M. and Lagarde, M. (2012). "Promoting Healthy Behaviours and Improving Health Outcomes in Low and Middle Income Countries: A Review of the Impact of Conditional Cash Transfer Programmes," *Preventive Medicine* 55(SUPPL.): S95–S105. Available at http://dx.doi.org/10.1016/j.ypmed.2011.11.015 (accessed January 3, 2017).

Rosenberg, T. (2016). "Labeling the Danger in Soda," *New York Times*. Available at http://opinionator.blogs.nytimes.com/2016/03/30/labeling-the-danger-in-soda/?smprod=nytcore-iphone&smid=nytcore-iphone-share&_r=0 (accessed January 3, 2017).

Shah, A.K., Mullainathan, S., and Shafir, E. (2012). "Some Consequences of Having Too Little," *Science* 338(6107): 682–685. Available at http://www.sciencemag.org/content/338/6107/682.abstract (accessed January 3, 2017).

Spears, D. (2011). "Economic Decision-Making in Poverty Depletes Behavioral Control," *The B.E. Journal of Economic Analysis & Policy*, 11(1).

Sturm, R., An, R., Segal, D., and Patel, D. (2013). "A Cash-Back Rebate Program for Healthy Food Purchases in South Africa: Results from Scanner Data," *American Journal of Preventive Medicine* 44(6): 567–572. Available at http://www.sciencedirect.com/science/article/pii/S0749379713001682 (accessed January 3, 2017).

Tarozzi, A., Mahajan, A., Yoong, J., and Blackburn, B. (2010). "Commitment Mechanisms and Compliance with Health-Protecting Behavior: Preliminary Evidence from Orissa (India)," *American Economic Review*, 99(2).

Thaler, R. and Sunstein, C.R. (2009). *Nudge: Improving Decisions About Health, Wealth and Happiness*. New York: Penguin.

WHO (2013). "Noncommunicable Diseases Fact Sheet," March 2013 Fact Sheet. Available at http://www.who.int/mediacentre/factsheets/fs355/en/ (accessed January 3, 2017).

PART IV

The role of providers, insurers, and government

11

APPLICATIONS OF BEHAVIORAL ECONOMICS TO CLINICAL QUALITY IMPROVEMENT

Daniella Meeker and Jason N. Doctor

Introduction

An enormous level of trust and respect is placed in the medical profession, which over the last three centuries has presided over the development of a scientific field devoted to knowledge of human physiology. While medical knowledge is still vastly incomplete, there is an even bigger and more pressing problem: Physicians very frequently fail to provide care that is supported by the evidence that has been developed. It is estimated that half of the care that patients actually receive is not based on the best available evidence (Jencks et al., 2003; McGlynn et al., 2003) and one-third of health expenditures do not improve health (Saunders and Smith, 2013). Perhaps because medical knowledge has a logical basis, physicians are by association seen as rational decision-makers themselves who, with the proper training, are ever mindful of scientific knowledge about best practices and appropriately apply this knowledge to each individual patient. However, physicians face constraints and influences that affect the quality of their decisions.

Clinicians have to make decisions under time pressure and they often rely on shortcuts in reasoning and, like other people, they may form habits. Learning from medical decisions is difficult because there is often no immediate consequence, or feedback from patients. An unnecessary radiograph can lead to cancer, and failure to prescribe a statin may result in a heart attack, but these events happen years from the time of the decision. In the clinic, repeated decisions may deplete the cognitive resources necessary for making good decisions (Dai et al., 2015; Linder et al., 2014; Owens, 2007; Reiner and Krupinski, 2012; Samkoff and Jacques, 1991). Interruption, which is common, can derail the deliberate thought processes of even the most organized and focused physician (Fordyce et al., 2003). Physicians are susceptible to cultural and social influences—organizational norms and systems, which may vary considerably from best practices, can have a strong impact over which diagnostic tests and treatments are chosen. While most physicians report that they are not influenced by marketing, evidence suggests otherwise (Spurling et al., 2010). Similarly, information systems intended to assist practitioners can be designed in ways that may inadvertently facilitate harmful orders. For example, one study found that a widely used physician order entry system facilitated 22 types of medication error risks (Koppel et al., 2005). Understanding the scientific underpinnings of decision science may be necessary to achieve the goals of improved care, reduced costs, and better population health.

In economics, modeling the human being as fully rational—so-called "*homo economicus*"—was a byproduct of other pursuits, notably the development of a theory of games (Von Neumann and Morgenstern, 2007). Rationality "greased the wheels" of an important enterprise that set out to identify strategies and policies for dealing with competitive situations. Similarly in building a healthcare system, with all the many components necessary to care for patients, it was expedient to assume that physicians are rational agents too. Yet, this assumption, which simplified the development of workflows, departments, services, and codification of medical knowledge bases and electronic health record systems, was rarely challenged until recently. The field of "behavioral economics," which incorporates insights from psychology to explain decision-making, has been the subject of growing interest. Much of the work in this area developed out of a body of literature that tested assumptions of *homo economicus* and then revised our understanding of decision-making to accommodate predictable irrationalities. Some of this work has focused on quality improvement in medicine and physician decision-making. To understand how behavioral economics is used, or can be used, in quality improvement, it helps to review current approaches that are implicitly based on the rational theory of choice.

Quality improvement interventions based on pay-for-performance designs assume a provider acts in accordance with *homo economicus*, whom in addition to his rationality, acts at least in part to further his own gain (Davis et al., 1995). Policymakers promoting pay-for-performance programs conjectured that providers just needed the right monetary incentive to improve performance. This implies providers are so rational that, despite a capability to provide optimal care, they simply are not doing so because it is not in their rational self-interest. With pay-for-performance, physicians are paid for meeting performance targets in the provision of care or in affecting patient outcomes. Findings indicate that responses to incentives are complex. For example, money can "crowd out" intrinsic motivation to do well (Frey and Oberholzer-Gee, 1997). This happens when personal reasons for performance are laid aside for money, resulting in changes to how we view our work. A physician asked to perform a best practice for added pay may be more likely to see herself as needing to meet a narrow incentive target than to provide good care overall. Such was the finding in one of the largest pay-for-performance studies, which was conducted in the United Kingdom (Campbell et al., 2009). These authors showed that while large incentives seemed to work on targeted measures, performance worsened on measures for which no incentive was given—an indication that pay-for-performance may not be the ideal solution. Thus strategies relying on pay-for-performance alone may not influence individual providers' decisions—leading to interest and innovation in alternative approaches.

Interventional studies notwithstanding, in practice, the "pay" in pay-for-performance does not easily reach physicians. A recent RAND report looked nationally at alternative payment models in the US and found nearly all physicians working within these alternative payment schemes did not receive even a dollar in direct payment (Friedberg et al., 2015). Payment overwhelmingly went to the healthcare organization or clinical practice, where it was spent on support staff or office equipment. Most physicians indicated that they were evaluated with respect to general productivity, which included providing low value services. In this environment, a physician might respond by acting as a "free rider"—someone who stays productive to gain high marks, but allows others to work toward the practice incentive from which s/he will also benefit if it is achieved. Further, a physician may not cooperate in achieving the group practice incentive even when it is in his or her best interest to do so.

While pay-for-performance has shortcomings, perhaps the most widely argued theory of behavioral economics' *negative* impact on medical decision-making is physicians' unconscious susceptibility to industry marketing. While physicians deny that they are influenced by marketing, (McKinney et al., 1990), this perception is contradicted by numerous findings showing that exposure to pharmaceutical campaigns influences prescribing decisions, including bias toward relatively ineffective and more expensive treatments (Wazana, 2000). Results suggest that this bias is unconscious—physicians may not remember the content of a free education course and even assert that campaigns will not bias their decisions, but subsequently favor the marketed product nonetheless (Spingarn et al., 1996; Steinman et al., 2001). In a cautionary commentary on the issue, Dana and Loewenstein (2003) argue that limiting the value of gifts is not sufficient to address unconscious bias—citing evidence from social science that recipients of even small gifts evaluate medical claims and interests of the gift givers in a more favorable light. The ineffectiveness of pay-for-performance stands in contrast to the effectiveness of marketing strategies that are not overtly "pay-for-prescribing," but are highly influential at promoting even comparatively ineffective medication orders.

Meanwhile, quality improvement alternatives to pay-for-performance have shown heterogeneous results and sensitivity to context (Kaplan et al., 2010; Lau et al., 2015; Oxman et al., 1995). Such quality improvement interventions have included educational strategies, both in traditional forms (classroom, paper-based materials) and in the form of "academic detailing" (Soumerai and Avorn, 1990)—more intimate outreach inspired by the success of detailing by pharmaceutical company representatives in influencing prescribing decisions. Detailing strategies include personal informational meetings with clinicians, a common practice of pharmaceutical sales representatives, but to encourage evidence-based delivery of care instead of product promotion. Engaging organizational opinion leaders in targeted campaigns is a related approach that similarly leverages social and professional context (Borbas et al., 2000). Personalized audit-and-feedback approaches have also been employed (Jamtvedt et al., 2006). These approaches include interventions where benchmarks from comparable organizations provide achievable targets (Kiefe et al., 2001), or working groups where results are discussed and compared (Vervloet et al., 2016). These approaches have demonstrated success in changing clinician behavior.

With the advent of electronic medical records, a diverse array of electronic alerts meant to remind physicians to order tests, manage medications or provide other health services have also been implemented and evaluated with equally diverse results (Bright et al., 2012). One meta-analysis showed that timely, specific electronic recommendations may be effective features of decision support systems (Kawamoto et al., 2005). Many operationally driven quality improvement programs go unevaluated, and underreporting is a threat to conclusions based on literature, but many published conclusions point to a role for success factors beyond merely supplying clinicians with information about best practices and the latest evidence. A recent review-of-reviews covering 91 systematic reviews of quality improvement interventions in primary care signaled that features such as tailoring and leveraging professional status and influence were key predictors of success in changing physician practices. While some pay-for-performance strategies were effective when incentives were large (and included some nudges like intermittency and unpredictability of rewards), generally, effect sizes were small (Lau et al., 2015). These results, taken together, suggest that it may be effective to approach clinical quality improvement using behavioral insights, or "nudges," particularly with interventions that appeal to social concerns and target habitual ordering behaviors that drive inappropriate care.

The term "nudge" is associated with a set of social and cognitive devices that persuade decisions in subtle ways while preserving choice. This approach is well matched to the goal of encouraging uptake of effective evidence-based treatments in healthcare. Nudges are non-intrusive persuaders such as different frames, default settings, feedback mechanisms, and social cues, which influence choice in a certain direction. Nudges are appealing in medicine because while they are designed to improve care decisions, they do not do so by limiting the choices available to physicians (Thaler and Sunstein, 2008).

Applying a behavioral economics framework: examples in overuse

The predominant business model for health care has been fee-for-service—that is, physicians are paid on the basis of services rendered, incentivizing volume of services delivered without necessarily incentivizing health outcomes. However, an increasing number of payers are creating payment models that incentivize value over volume—measured by either health outcomes or adherence to guidelines. The deep-rooted habits of fee-for-service systems will need to be changed. Advancing evidence-based care by influencing clinical decision-making requires applying decision science to clinical contexts in ways that change habits that are a result of the structure of health care delivery. Nudges can be applied to both habitual overuse and underuse. For example, tools like automated standing orders for appropriate services can dramatically impact underuse by eliminating psychological frictions to ordering behavior (Dexter et al., 2004). However, reducing unnecessary services requires either preventing orders that are often placed habitually, or overcoming the forward momentum of an order that has already been placed. That is, the forces that make underuse a relatively easy target for physician decision-making are exactly those that make overuse challenging.

In this section we discuss several randomized experiments of applications of behavioral economics to curtail inappropriate prescribing of antibiotics. Antibiotic overprescribing is representative of several quality improvement concerns that can be addressed with nudges. Antibiotics represent a major public health issue that has been difficult to address and which persists as a problem. We focus on "supply-side" behavioral interventions—those that target clinicians to change prescribing behavior. Antibiotic ordering decisions have proven to be among the most challenging physician behaviors to change. Despite consensus on guidelines and national campaigns devoted to education and performance measurement, inappropriate prescribing rates for acute respiratory infections still hover around 50 percent in the United States. Thus, antibiotic prescribing decisions are a prime target for introducing nudges that improve guideline concordance. Each intervention we describe leveraged principles of behavioral economics that have been successfully applied in other contexts: *public commitment*, *choice architecture*, *defaults*, *social norms*, and *peer accountability*. For each type of nudge, we describe the rationale behind the design decisions and the experimental results.

Public commitments

Many, if not all, physicians support quality improvement but most have difficulty making real improvements at the point-of-care in their own practices. Physicians that aspire to improve their performance, or model good practices to others, might benefit from public pledges or commitments—statements of dedication to a course of action that are visible to others. People place a high value on consistency, generally act consistent with previous commitments, and follow through with their public commitments to avoid disapproval by their peers. Public commitment has been shown to increase recycling, heighten participation in conservation programs, boost monetary contributions to charities, enhance the likelihood of voting in an upcoming election and lead dieters to lose more weight (Baca-Motes et al., 2013; Burn and Oskamp, 1986; Cialdini, 2010; Cialdini et al., 1978; De Leon and Fuqua, 1995; Deutsch and Gerard, 1955; Krendl et al., 1992; Nyer and Dellande, 2010). Publicly committing to a course of action also influences people to later justify the importance of that action and identify the action with their self-image, which may enhance the effect of commitment.

In a randomized trial, we studied the effect of physicians committing publicly to using antibiotics only when they were absolutely necessary (Meeker et al., 2014). Physicians in the experimental condition signed a letter pledging judicious use of antibiotics, then a poster of that pledge was displayed on the exam room wall, displaying the clinician's signature and photograph. The rate of prescribing antibiotics for viral respiratory infections dropped nearly 20 percentage points compared to the control group, which had no poster. A number of factors may have played into the success of this public commitment. Not only may the poster have compelled physicians to honor their commitment, but also it may have provided them with "cover" from real or perceived patient demands. A physician may be less concerned about satisfaction ratings going down when they refuse to prescribe antibiotics in front of a poster that clearly states their personal prior commitment to reducing antibiotic use. In general, public commitments may offer a powerful approach to change physician behavior because commitments are respected, and likely patient beliefs are that commitments promote desirable healthcare practices.

Socially motivated performance reporting

Previous research demonstrates that comparisons of a person's behavior to others' appropriate behavior is more effective than modest economic incentives or self-interest (Heyman and Ariely, 2004; Nolan et al., 2008). This is particularly true when the comparisons are to similar associates (Goldstein et al., 2008; Hornstein et al., 1968; Murray et al., 1984; White et al., 2002). An early field experiment showed that informing households of the average neighborhood energy consumption (a "descriptive social norm") led to reduced consumption among those previously consuming more than average. However, this had the unintended consequence of increasing consumption for some of the below-average consumers. Adding a message conveying social approval for desirable behavior (an "injunctive norm") prevented those below-average consumers from increasing their energy consumption (Schultz et al., 2007).

Peer comparison coupled with audit-and-feedback is a well-established quality improvement strategy (Gerber et al., 2013; Jamtvedt et al., 2006; Kiefe et al., 2001; Ward et al., 2014). However, mixed results may be due to subtleties like those discovered in the energy consumption study. In a recent randomized trial, we produced a variant on the traditional approach, where the peer comparison intervention involved calculating the proportion of inappropriate prescriptions for each clinician from the electronic health record on a monthly basis, and ranking clinicians from the lowest to the highest inappropriate prescribing rate within each region; we also provided an injunctive norm of approval to the top performers without revealing their place on the distribution ("You are a top performer"). This approach almost completely eliminated inappropriate prescribing to a rate of 3.7 percent—an 81 percent reduction (Meeker et al., 2016). This type of peer comparison performance reporting is highly scalable if performance can be measured electronically, requiring no changes to clinical workflow or the electronic health record system. All that is needed is a well-specified electronic performance measure such as those endorsed by the National Quality Forum and access to the data system to produce peer comparisons and email or US mail to deliver reports. In addition, peer comparison is applicable to any over- or underuse problem in medicine that is measurable.

Accountable justifications

There is evidence that accountability improves decision-making accuracy; public justification also engenders reputational concerns aligning decisions with best practices (Baumeister and Hutton, 1987; Jerdee and Rosen, 1974; Lerner and Tetlock, 1999; Milinski et al., 2002). These same principles may be applied within the electronic medical record. In an electronic health record-integrated intervention, clinicians were asked to explicitly justify non-guideline-concordant antibiotic prescribing through a computerized prompt. While this type of intervention has been implemented in the past, the notes were not readily visible to peers, which eliminated accountability to an audience (Litzelman, 1993). By adding a separate "Antibiotic Justification" note in each patient's electronic record, rationale for divergence from guidelines was viewable by other clinicians. If no justification was entered, "no justification given" appeared in the placeholder section. Encounters could not be closed without the clinician's acknowledgment of the prompt. Accountable justification can be used to address any quality improvement problem for which there are established best practices, because physicians should always be able to justify their actions in such cases. By invoking professional accountability, this intervention reduced inappropriate prescribing from 23.2 percent to 5.2 percent (Meeker et al., 2016). These interventions are relatively easy to scale in electronic medical records, and while they require encoding guidelines directly into organization-specific information systems and implementing real-time prompts, larger organizations may benefit from these types of approaches to improving care.

Choice presentation

There is an extensive body of evidence supporting the effect of the choice presentation format on preference. This ranges from grocery store shelf placement (Dreze et al., 1995) to election ballots (Miller and Krosnick, 1998) and survey choices (Krosnick and Alwin, 1987). From a rational perspective of classical economics, as long as the description is accurate, the manner in which options are described should not affect preference. The fundamental observations, though, contradict this invariance assertion. Research finds that: (1) salient alternatives will more likely be chosen over less salient ones; (2) people infer that an alternative presented is one that ought to be considered, thus reducing the likelihood that an inappropriate option will be chosen; and (3) when offered multiple selections, individuals tend to allocate selections evenly over grouped options ("diversification bias") (Read and Loewenstein, 1995). In studies that have compared novices to experts, the effects are greater on novices, but experts are not immune to these effects (Fox et al., 2005).

The impact of choice architecture is robust and has been replicated in numerous scenarios, including by the most tenacious adherents to the rational perspective (Grether and Plott, 1979). While many of the studies of this phenomenon have been conducted in controlled laboratory experiments in hypothetical examples, the effects of choice presentation are manifest in many situations with real-world consequences.

With the advent of electronic ordering systems, the influence of choice presentation in electronic order sets on medical ordering has received significant attention (Ash et al., 2007; Campbell et al., 2006; Sengstack, 2010; Walsh et al., 2006). While much of the focus has been on patient safety, there is potential to use choice presentation to nudge clinicians in the direction of a better option while preserving accessibility of alternatives. For example, when antibiotics are indicated for an acute respiratory infection, it is preferable for antibiotic stewardship to prescribe narrow-spectrum antibiotics over broad-spectrum antibiotics, because such antibiotics are less likely to promote resistant bacteria. In a randomized analog trial, primary care physicians were presented vignettes of three different cases of acute respiratory infections that warranted antibiotic prescriptions. After each case, each physician was randomly assigned to one of two versions of hypothetical order sets and asked to select a prescription. In one version, all broad-spectrum antibiotics were presented in a single row, with one row for each narrow-spectrum choice (devoting more screen space to the narrow-spectrum choices and less screen space to broad-spectrum choices). The other condition reversed the grouping. When appropriate choices were made more salient, clinicians were significantly more likely to select those choices. Perhaps telling of limitations of nudges, the same study also encouraged over-the-counter treatments instead of antibiotics, but the effect size was much smaller (7.6 percentage point reduction vs. 14.4) (Tannenbaum et al., 2015). These results suggest that design of choice presentation in order entry systems, an area that has received substantial attention from safety experts, might benefit from applying the extensive body of evidence regarding choice presentation results identified in other disciplines.

Defaults

Defaults are options that are exercised if the decision-maker takes no special action to opt-out. Classical economics holds that choices are not affected by which option is designated the default. In contrast, a number of behavioral studies have demonstrated a surprisingly strong bias toward choosing the default option in areas as diverse as organ donation (Johnson and Goldstein, 2003), 401(k) investment (Madrian and Shea, 2001), and the prescription of generic over-branded drugs (Patel et al., 2014). Defaults may affect behavior for a number of reasons: 1) they may be perceived as a recommended action; 2) they require less effort; 3) they serve as a reference point so that relative disadvantages of alternatives are viewed as losses that loom larger than relative advantages; and 4) they may seem less anxiety-provoking as people tend to regret active choices that lead to poor outcomes more than they regret failures to act that lead to poor outcomes.

Defaults may have both positive and negative consequences in care. As described above, standing orders for vaccines are effective in increasing adherence to recommended services (Dexter et al., 2004). At the same time, default settings on order entry tools may create errors in dosages that can cause serious harm (Koppel et al., 2005; Walsh et al., 2006). "Smart" defaults that vary dose by patient characteristics could overcome this limitation (Fischer et al., 2008).

Related to defaults are enhanced "active choice' conditions (Keller et al., 2011). These conditions require a choice be made and they emphasize the negative characteristics associated with the less desirable alternative. A question remains, however, about whether overuse can be discouraged through active choice. In a randomized trial of active choice prescription orders, non-antibiotic alternatives were recommended and emphasized prominently for all patients with upper respiratory infections; antibiotics were discouraged for these qualifying visits. In this case, there was a trend toward significance but no statistically significant difference between those given and not given the active choice (Meeker et al., 2016). There are a number of possible explanations for this. These types of decision support used to deliver the active choice may not fit well into workflow and potentially conflict with other behavioral tendencies—for example, if the clinician enters prescription orders after assigning a diagnosis and verbalizing a care plan that includes antibiotics, changing to a non-antibiotic order might conflict with a commitment that he has made to a patient. Even if a decision has not been verbalized, if the clinician has already developed a mental implementation plan for placing an antibiotic order, she will be more likely to adhere to that plan (Gollwitzer, 1999). More research may be needed regarding how and under what circumstances defaults might be employed to curtail habitual overuse.

Limitations and precautions

There are a number of considerations that practitioners of behavioral economics strategies for quality improvement might note. Most of the nudges above are "System 1 nudges" that leverage non-deliberative, largely unconscious psychological tendencies. There is evidence that people prefer "System 2 nudges"—deliberate processing—because they respect individual agency to a greater degree (Sunstein, 2016). This preference may be particularly salient among professionals who are valued for their deliberative abilities, potentially resulting in defiance or rationalization driven by self-serving bias (Dana and Loewenstein, 2003). In other cases, there may be unintended consequences or backlash effects that should be monitored. For example, public commitments described above may carry the risk of operating in similar ways to public conflict-of-interest disclosures. Studies comparing individuals assigned to conflict-of-interest conditions found that disclosing conflicts may potentially give advisors license to engage in the very activity it is designed to discourage (Cain et al., 2005).

The use of defaults should also be practiced judiciously—there is evidence that opt-in and opt-out defaults in electronic order sets can lead to an increase in omission and commission errors, respectively (Ansher et al., 2014). The findings suggest these errors have only a mild effect on outcomes, but defaults should be carefully designed to be as patient-centered as possible, taking into account allergies, contraindications, and other safety risks. As described above, peer comparisons have been shown to be effective in a number of contexts, but describing performance norms if typical performance is poor may unconsciously encourage high performers to regress to a less desirable, more normative level of performance. Mitigating strategies should be considered—including leveraging injunctive norms—signaling to high performers that peers approve and aspire to similar status, even if peers have not obtained the performance goals themselves (Göckeritz et al., 2010). The intricacy of clinical habits, workflows, professional relationships, and systemic influences provide a number of different channels and challenges for design and implementation of interventions inspired by behavioral economics theory.

Conclusion

Future directions

There is no shortage of opportunities to apply principles of behavioral economics and decision science to improve quality, efficiency, and safety. Here we raise two opportunities that exemplify future possibilities—the first to curtail overuse, the second to improve overall quality and safety, and reduce disparities.

One particularly salient and complex example of overuse is opioid prescribing—a phenomenon that has reached crisis proportions in the United States—where 46 Americans die each day from prescription opioid overdoses; a death every 30 minutes (Centers for Disease Control, 2014). In some ways the current opioid crisis might be traced to social influence, exerted by specialist physicians, which resulted in supply-side expansion—the greater supply of drugs through prescribing. As noted by Von Korff (2012, p. 303):

> In the 1990s, palliative care specialists defined opioid prescribing standards for chronic non-cancer pain. These guidelines, based almost entirely on expert opinion, were then generalized to primary care practice without research, evaluation, or meaningful input from primary care physicians.

Regrettably, this expansion was without a scientific foundation in clinical research (American Pain Society, 1997).

In the medical community, other social influences may have played a role in expanded use. Widely cited perspectives tied a physician's self-image to their efforts to rescue patients from pain. "To allow a patient to experience unbearable pain or suffering is unethical medical practice," stated one influential paper (Wanzer et al., 1989). Opioid prescribing in non-cancer pain leads to adverse selection—those who take the highest doses and for the longest periods of time are naturally the persons most prone to addiction and overdose death (Sullivan et al., 2005). Poor pain control as characterized by breakthrough pain requiring higher doses of these drugs is also a frequent problem (Narayana et al., 2015). The physiology and psychology of pain and addiction have been tightly coupled with the decision psychology of prescription orders.

Just as social devices resulted in rapid prescribing expansion, so may they also be employed to reverse the expansion. Interventions that reduce the general environmental availability of prescription narcotics will require changing social norms, public commitments to curtail poisonings and uses of default options, in the electronic health record, that give injunctions to begin tapering use. With private benefits for chronic pain largely unsubstantiated and social costs pronounced, encouraging judicious use has a clear ethical justification. Given the extent to which peer influence and social norms contributed to the opioid crisis, these nudges may also be effective in reversing this distressing trend.

Other future opportunities may come from addressing the ways in which physicians' performance is impacted by cognitive load. Most of the pursuits described above have focused on interventions that are intended to change physicians' habitual behaviors with extrinsic nudges. Emerging evidence suggests that quality of decisions can erode as effortful judgments accumulate over the course of a session or shift, and decision-makers revert to habit over reasoning. This phenomenon, coined *decision fatigue*, has been shown to impact consumers and professionals ranging from judges to clinicians (Govorun and Payne, 2006; Hagger and Chatzisarantis, 2013; Linder et al., 2014; Masicampo and Baumeister, 2008; Schmeichel et al., 2003; Weinshall-Margel and Shapard, 2011). This reversion to (often bad) habits is linked in psychological literature with concepts of *ego-depletion*, willpower and self-control, but these constructs differ, as some habits may not be consciously controlled. Troublingly, decision fatigue has been shown to surface unconscious racial biases that are prevalent among clinicians as well as the general population (Govorun and Payne, 2006; Ma et al., 2013; Van Ryn, 2002). These findings raise the question of whether healthcare disparities with sensitivity to providers' actions could be aggravated by cognitive load (Burgess, 2010). Given the implications, decision fatigue may be a particular area of concern for future research in policy and quality improvement.

Encouraging experimental results have emerged that may suggest viable interventions to counteract decision fatigue, but not without controversy. Fast-acting physiological interventions may help address the deficits brought on by decision fatigue by enhancing self-control. Investigations have shown that administering glucose during an experimental session can reduce the symptoms of decision fatigue on delay discounting tasks that test subjects' abilities to forgo short-term rewards in favor of larger long-term rewards (Gailliot and Baumeister, 2007; Gailliot et al., 2007; Masicampo and Baumeister, 2008). More recent findings have shown that these results can be obtained simply with sweet mouth-tastes without requiring glucose ingestion (Hagger and Chatzisarantis, 2013). Importantly, these results were not replicated in rigorously designed studies by other scientists (Lange and Eggert, 2014)—raising the question of whether alternative interventions devoted to self-reinforcing practices (Dewitte et al., 2009) might be more difficult to implement than providing physicians a mid-shift sugar boost, but ultimately more effective. Future research devoted to improving clinical decision-making under cognitively demanding circumstances may benefit from these and other findings in decision science.

Summary

Every decision a clinician makes has the potential to promote efficient, evidence-based care and contribute to population health. Some of these decisions are deliberative and carefully considered, but time and resource constraints require that most clinical decisions be made quickly and without circumspection. In these cases, applied behavioral economics has the potential to influence decisions toward best practices. Policy changes that aim at redesigning the health system to improve care often fail to consider components of decision processes that may have at their origins structural and systemic factors, but have become ingrained as part of cultural and individual habits. Small, scalable changes based on the science of decision-making have the potential to promote evidence-based care, reduce overuse, and increase compliance with best practices. Future work might investigate how similar interventions can reduce racial and ethnic disparities. Public commitments, default orders, providing clinicians with feedback on their performance relative to their top-performing peers, requiring physicians to justify their actions in an observable way, and even physiological interventions, may be effective in improving quality of care. The idea that small changes to the clinic environment can have a large impact on clinician prescribing behavior is a practical approach to quality improvement. Better care in the future may depend on harnessing the power of decision science to improve population health (McGlynn, 1998; Orszag, 2008).

References

American Pain Society (1997). "The Use of Opioids for the Treatment of Chronic Pain. A Consensus Statement from the American Academy of Pain Medicine and the American Pain Society," *Clinical Journal of Pain* 13(1): 6–8.

Ansher, C., Ariely, D., Nagler, A., Rudd, M., Schwartz, J., and Shah, A. (2014). "Better Medicine by Default," *Medical Decision Making* 34(2): 147–158.

Ash, J.S., Sittig, D.F., Poon, E.G., Guappone, K., Campbell, E., and Dykstra, R.H. (2007). "The Extent and Importance of Unintended Consequences Related to Computerized Provider Order Entry," *Journal of the American Medical Informatics Association* 14(4): 415–423.

Baca-Motes, K., Brown, A., Gneezy, A., Keenan, E.A., and Nelson, K.D. (2013). "Commitment and Behavior Change: Evidence from the Field," *Journal of Consumer Research* 39(5): 1070–1084.

Baumeister, R.E. and Hutton, D.G. (eds) (1987). *Self-Presentation Theory: Self-Construction and Audience Pleading*. Theories of Group Behavior. New York: Springer Verlag.

Borbas, C., Morris, N., McLaughlin, B., Asinger, R., and Gobel, F. (2000). "The Role of Clinical Opinion Leaders in Guideline Implementation and Quality Improvement," *Chest Journal* 118(2_suppl): 24S–32S.

Bright, T.J., Wong, A., Dhurjati, R., Bristow, E., Bastian, L., Coeytaux, R.R., Samsa, G., Hasselblad, V., Williams, J.W., Musty, M.D., Wing, L., Kendrick, A.S., Sanders, G.D., and Lobach, D. (2012). "Effect of Clinical Decision-Support Systems: A Systematic Review," *Annals of Internal Medicine* 157(1): 29–43.

Burgess, D.J. (2010). "Are Providers More Likely to Contribute to Healthcare Disparities under High Levels of Cognitive Load? How Features of the Healthcare Setting May Lead to Biases in Medical Decision Making," *Medical Decision Making* 30(2): 246–257.

Burn, S.M. and Oskamp, S. (1986). "Increasing Community Recycling with Persuasive Communication and Public Commitment," *Journal of Applied Social Psychology* 16(1): 29–41.

Cain, D.M., Loewenstein, G., and Moore, D.A. (2005). "The Dirt on Coming Clean: Perverse Effects of Disclosing Conflicts of Interest," *The Journal of Legal Studies* 34(1): 1–25.

Campbell, S., Reeves, D., Kontopantelis, E., Sibbald, B., and Martin Roland, D.M. (2009). "Effects of Pay for Performance on the Quality of Primary Care in England," *New England Journal of Medicine* 361: 368–378.

Campbell, E.M., Sittig, D.F., Ash, J.S., Guappone, K.P., and Dykstra, R.H. (2006). "Types of Unintended Consequences Related to Computerized Provider Order Entry," *Journal of the American Medical Informatics Association* 13(5): 547–556.

Centers for Disease Control (2014). "CDC Vital Signs," July 2014. Available at http://www.cdc.gov/Vitalsigns/Opioid-Prescribing/ (accessed December 31, 2016).

Cialdini, R.B. (2010). *Influence: Science and Practice*. 5th edition. Boston: Allyn & Bacon.

Cialdini, R.B., Cacioppo, J.T., Bassett, R., and Miller, J.A. (1978). "Low-Ball Procedure for Producing Compliance: Commitment Then Cost," *Journal of Personality and Social Psychology* 36(5): 463.

Dai, H., Milkman, K.L., Hofmann, D.A., and Staats, B.R. (2015). "The Impact of Time at Work and Time Off from Work on Rule Compliance: The Case of Hand Hygiene in Health Care," *Journal of Applied Psychology* 100(3): 846.

Dana, J. and Loewenstein, G. (2003). "A Social Science Perspective on Gifts to Physicians from Industry," *The Journal of the American Medical Association (JAMA)* 290(2): 252–255.

Davis, D., Thomson, M.A., Oxman, A.D., and Haynes, R.B. (1995). "Changing Physician Performance: A Systematic Review of the Effect of Continuing Medical Education Strategies," *JAMA* 274(9): 700–705.

De Leon, I.G. and Fuqua, R.W. (1995). "The Effects of Public Commitment and Group Feedback on Curbside Recycling," *Environment and Behavior* 27: 233–250.

Deutsch, M. and Gerard, H.B. (1955). "A Study of Normative and Informative Social Influences Upon Individual Judgment," *Journal of Abnormal Social Psychology* 51: 629–636.

Dewitte, S., Bruyneel, S., and Geyskens, K. (2009). "Self-Regulating Enhances Self-Regulation in Subsequent Consumer Decisions Involving Similar Response Conflicts," *Journal of Consumer Research* 36(3): 394–405.

Dexter, P.R., Perkins, S.M., Maharry, K.S., Jones, K., and McDonald, C.J. (2004). "Inpatient Computer-Based Standing Orders vs Physician Reminders to Increase Influenza and Pneumococcal Vaccination Rates: A Randomized Trial," *JAMA* 292(19): 2366–2371.

Dreze, X., Hoch, S.J., and Purk, M.E. (1995). "Shelf Management and Space Elasticity," *Journal of Retailing* 70(4): 301–326.

Fischer, M.A., Vogeli, C., Stedman, M., Ferris, T., Brookhart, M.A., and Weissman, J.S. (2008). "Effect of Electronic Prescribing with Formulary Decision Support on Medication Use and Cost," *Archives of Internal Medicine* 168(22): 2433–2439.

Fordyce, J., Blank, F.S., Pekow, P., Smithline, H.A., Ritter, G., Gehlbach, S., Benjamin, E., and Henneman, P.L. (2003). "Errors in a Busy Emergency Department," *Annals of Emergency Medicine* 42(3): 324–333.

Fox, C.R., Ratner, R.K., and Lieb, D.S. (2005). "How Subjective Grouping of Options Influences Choice and Allocation: Diversification Bias and the Phenomenon of Partition Dependence," *Journal of Experimental Psychology: General* 134(4): 538.

Frey, B. and Oberholzer-Gee, F. (1997). "The Cost of Price Incentives: An Empirical Analysis of Motivation Crowding-Out," *The American Economic Review* (87)4: 746–755.

Friedberg, M., Chen, P.G., White, C., Jung, O., Raaen, L., Hirshman, S, ... and Lipinski, L. (2015). "Effects of Health Care Payment Models on Physician Practice in the United States," *RAND Corporation*.

Gailliot, M.T. and Baumeister, R.F. (2007). "The Physiology of Willpower: Linking Blood Glucose to Self-Control," *Personality and Social Psychology Review* 11(4): 303–327.

Gailliot, M.T., Baumeister, R.F., DeWall, C.N., Maner, J.K., Plant, E.A., Tice, D.M., Brewer, L.E., and Schmeichel, B.J. (2007). "Self-Control Relies on Glucose as a Limited Energy Source: Willpower Is More Than a Metaphor," *Journal of Personality and Social Psychology* 92(2): 325.

Gerber, J.S., Prasad, P.A., Fiks, A.G. et al. (2013). "Effect of an Outpatient Antimicrobial Stewardship Intervention on Broad-Spectrum Antibiotic Prescribing by Primary Care Pediatricians: A Randomized Trial," *JAMA* 309(22): 2345–2352.

Göckeritz, S., Schultz, P., Rendón, T., Cialdini, R.B., Goldstein, N.J., and Griskevicius, V. (2010). "Descriptive Normative Beliefs and Conservation Behavior: The Moderating Roles of Personal Involvement and Injunctive Normative Beliefs," *European Journal of Social Psychology* 40(3): 514–523.

Goldstein, N.J., Cialdini, R.B., and Griskevicius, V. (2008). "A Room with a Viewpoint: Using Social Norms to Motivate Environmental Conservation in Hotels," *Journal of Consumer Research* 35: 472–482.

Gollwitzer, P.M. (1999). "Implementation Intentions: Strong Effects of Simple Plans," *American Psychologist* 54(7): 493.

Govorun, O. and Payne, B.K. (2006). "Ego-Depletion and Prejudice: Separating Automatic and Controlled Components," *Social Cognition* 24(2): 111–136.

Grether, D.M. and Plott, C.R. (1979). "Economic Theory of Choice and the Preference Reversal Phenomenon," *The American Economic Review* 69(4): 623–638.

Hagger, M.S. and Chatzisarantis, N.L. (2013). "The Sweet Taste of Success the Presence of Glucose in the Oral Cavity Moderates the Depletion of Self-Control Resources," *Personality and Social Psychology Bulletin* 39(1): 28–42.

Heyman, J. and Ariely, D. (2004). "Effort for Payment: A Tale of Two Markets," *Psychological Science* 15: 787–793.

Hornstein, H., Fisch, E., and Holmes, M. (1968). "Influence of a Model's Feelings About His Behavior and His Relevance as a Comparison Other on Observers' Helping Behavior," *Journal of Personality and Social Psychology* 10: 220–226.

Jamtvedt, G., Young, J.M., Kristoffersen, D.T., O'Brien, M.A., and Oxman, A.D. (2006). "Audit and Feedback: Effects on Professional Practice and Health Care Outcomes," *Cochrane Database of Systematic Reviews* 2(2).

Jencks, S., Huff, E.D., and Cuerdon, T. (2003). "Change in the Quality of Care Delivered to Medicare Beneficiaries, 1998–1999 to 2000–2001," *Journal of the American Medical Association* 289(3): 305–312.

Jerdee, T. and Rosen, B. (1974). "Effects of Opportunity to Communicate and Visibility of Individual Decisions on Behavior in the Common Interest," *J App Psych* 59: 712–716.

Johnson, E.J. and Goldstein, D.G. (2003). "Do Defaults Save Lives?" *Science* 302: 1338–1339.

Kaplan, H.C., Brady, P.W., Dritz, M.C., Hooper, D.K., Linam, W., Froehle, C.M., and Margolis, P. (2010). "The Influence of Context on Quality Improvement Success in Health Care: A Systematic Review of the Literature," *Milbank Quarterly* 88(4): 500–559.

Kawamoto, K., Houlihan, C.A., Balas, E.A., and Lobach, D.F. (2005). "Improving Clinical Practice Using Clinical Decision Support Systems: A Systematic Review of Trials to Identify Features Critical to Success," *BMJ* 330(7494): 765.

Keller, P.A., Harlam, B., Loewenstein, G., and Volpp, K.G. (2011). "Enhanced Active Choice: A New Method to Motivate Behavior Change," *Journal of Consumer Psychology* 21(4): 376–383.

Kiefe, C.I., Allison, J.J., Williams, O.D., Person, S.D., Weaver, M.T., and Weissman, N.W. (2001). "Improving Quality Improvement Using Achievable Benchmarks for Physician Feedback," *JAMA* 285(22): 2871–2879.

Koppel, R., Metlay, J.P., Cohen, A., Abaluck, B., Localio, A.R., Kimmel, S.E., and Strom, B.L. (2005). "Role of Computerized Physician Order Entry Systems in Facilitating Medication Errors," *JAMA* 293(10): 1197–1203.

Krendl, K.A., Olson, B., and Burke, R. (1992). "Preparing for the Environmental Decade: A Field Experiment on Recycling Behavior," *Journal of Applied Communication Research* 20(1): 19–36.

Krosnick, J.A. and Alwin, D.F. (1987). "An Evaluation of a Cognitive Theory of Response-Order Effects in Survey Measurement," *Public Opinion Quarterly* 51(2): 201–219.

Lange, F. and Eggert, F. (2014). "Sweet Delusion. Glucose Drinks Fail to Counteract Ego Depletion," *Appetite* 75: 54–63.

Lau, R., Stevenson, F., Ong, B.N., Dziedzic, K., Treweek, S., Eldridge, S., Everitt, H., Kennedy, A., Qureshi, N., Rogers, A., Peacock, R., and Murray, E. (2015). "Achieving Change in Primary Care-Effectiveness of Strategies for Improving Implementation of Complex Interventions: Systematic Review of Reviews," *BMJ Open* 5(12): e009993.

Lerner, J. and Tetlock, P. (1999). "Accounting for the Effects of Accountability," *Psychological Bulletin* 125(2): 255.

Linder, J.A., Doctor, J.N., Friedberg, M.W., Reyes Nieva, H., Birks, C., Meeker, D., and Fox, C. R. (2014). "Time of Day and the Decision to Prescribe Antibiotics," *JAMA Internal Medicine* 174(12): 2029–2031.

Litzelman, D., Dittus, R.S., Miller, M.E., and Tierney, W.M. (1993). "Requiring Physicians to Respond to Computerized Reminders Improves Their Compliance with Preventive Care Protocols," *Journal of General Internal Medicine* 8(6): 311–317.

Ma, D.S., Correll, J., Wittenbrink, B., Bar-Anan, Y., Sriram, N., and Nosek, B.A. (2013). "When Fatigue Turns Deadly: The Association between Fatigue and Racial Bias in the Decision to Shoot," *Basic and Applied Social Psychology* 35(6): 515–524.

Madrian, B.C. and Shea, D.F. (2001). "The Power of Suggestion: Inertia in 401(K) Participation and Savings Behavior," *The Quarterly Journal of Economics*, 116(4): 1149–1187.

Masicampo, E.J. and Baumeister, R.F. (2008). "Toward a Physiology of Dual-Process Reasoning and Judgment: Lemonade, Willpower, and Expensive Rule-Based Analysis," *Psychological Science* 19(3): 255–260.

McGlynn, E. (1998). "Assessing the Appropriateness of Care: How Much Is Too Much?" Rand Research Brief (Santa Monica, CA: Rand). Available at http://www.rand.org/pubs/research_briefs/RB4522.html (accessed December 31, 2016).

McGlynn, E.A., Asch, S.M., Adams, J., Keesey, J., Hicks, J., DeCristofaro, A., and Kerr, E.A. (2003). "The Quality of Health Care Delivered to Adults in the United States," *New England Journal of Medicine* 348(26): 2635–2645.

McKinney, W.P., Schiedermayer, D.L., Lurie, N., Simpson, D.E., Goodman, J.L., and Rich, E.C. (1990). "Attitudes of Internal Medicine Faculty and Residents Toward Professional Interaction with Pharmaceutical Sales Representatives," *JAMA* 264(13): 1693–1697.

Meeker, D., Knight, T.K., Friedberg, M.W. et al. (2014). "Nudging Guideline-Concordant Antibiotic Prescribing: A Randomized Clinical Trial," *JAMA Internal Medicine* 174(3): 425–431.

Meeker, D., Linder, J.A., Fox, C.R. et al. (2016). "Effect of Behavioral Interventions on Inappropriate Antibiotic Prescribing among Primary Care Practices: A Randomized Clinical Trial," *JAMA* 315(6): 562–570.

Milinski, M., Semmann, D., and Krambeck, H. (2002). "Reputation Helps Solve the 'Tragedy of the Commons,'" *Nature* 415: 424–426.

Miller, J.M. and Krosnick, J.A. (1998). "The Impact of Candidate Name Order on Election Outcomes," *Public Opinion Quarterly* 62(3, Autumn): 291–330.

Murray, D.M., Luepker, R.V., Johnson, A.C., and Mittelmark, M.B. (1984). "The Prevention of Cigarette Smoking in Children: A Comparison of Four Strategies," *Journal of Applied Social Psychology* 14: 274–288.

Narayana, A., Katz, N., Shillington, A.C., Stephenson, J.J., Harshaw, Q., Frye, C.B., and Portenoy, R.K. (2015). "National Breakthrough Pain Study: Prevalence, Characteristics, and Associations with Health Outcomes," *Pain* 156(2): 252–259.

Nolan, J.P., Schultz, P.W., Cialdini, R.B., Goldstein, N.J., and Griskevicius, V. (2008). "Normative Social Influence Is Underdetected," *Personality and Social Psychology Bulletin* 34: 913–923.

Nyer, P.U. and Dellande, S. (2010). "Public Commitment as a Motivator for Weight Loss," *Psychology & Marketing* 27(1): 1–12.

Orszag, P.R. (2008). "The Overuse, Underuse, and Misuse of Health Care," Statement before the Committee on Finance, US Senate, July 17.

Owens, J.A. (2007). "Sleep Loss and Fatigue in Healthcare Professionals," *The Journal of Perinatal & Neonatal Nursing* 21(2): 92–100.

Oxman, A., Thomson, M.A., Davis, D.A., and Haynes, R.B. (1995). "No Magic Bullets: A Systematic Review of 102 Trials of Interventions to Improve Professional Practice," *CMAJ: Canadian Medical Association Journal* 153(10): 1423.

Patel, M., Day, S., Small, D.S., Howell, J.T., Lautenbach, G.L., Nierman, E.H., and Volpp, K.G. (2014). "Using Default Options within the Electronic Health Record to Increase the Prescribing of Generic-Equivalent Medications: A Quasi-Experimental Study," *Annals of Internal Medicine* 161(10): S44–S52.

Read, D. and Loewenstein, G. (1995). "Diversification Bias: Explaining the Discrepancy in Variety Seeking between Combined and Separated Choices," *Journal of Experimental Psychology: Applied* 1(1): 34.

Reiner, B.I. and Krupinski, E. (2012). "The Insidious Problem of Fatigue in Medical Imaging Practice," *Journal of Digital Imaging* 25(1): 3–6.

Samkoff, J.S. and Jacques, C. (1991). "A Review of Studies Concerning Effects of Sleep Deprivation and Fatigue on Residents' Performance," *Academic Medicine* 66(11): 687–693.

Saunders, R. and Smith, M.D. (2013). "The Path to Continuously Learning Health Care," *Issues in Science and Technology* 29(3): 27–36.

Schmeichel, B.J., Vohs, K.D., and Baumeister, R.F. (2003). "Intellectual Performance and Ego Depletion: Role of the Self in Logical Reasoning and Other Information Processing," *Journal of Personality and Social Psychology* 85(1): 33.

Schultz, P.W., Nolan, J.M., Cialdini, R.B., Goldstein, N.J., and Griskevicius, V. (2007). "The Constructive, Destructive, and Reconstructive Power of Social Norms," *Psychological Science* 18(5): 429–434.

Sengstack, P. (2010). "Cpoe Configuration to Reduce Medication Errors," *Journal of Healthcare Information Management* 24(4).

Soumerai, S.B. and Avorn, J. (1990). "Principles of Educational Outreach ('Academic Detailing') to Improve Clinical Decision-Making," *JAMA* 263(4): 549–556.

Spingarn, R.W., Berlin, J.A., and Strom, B.L. (1996). "When Pharmaceutical Manufacturers' Employees Present Grand Rounds, What Do Residents Remember?" *Academic Medicine* 71(1): 86–88.

Spurling, G.K., Mansfield, P.R., Montgomery, B.D., Lexchin, J., Doust, J., Othman, N., and Vitry, A.I. (2010). "Information from Pharmaceutical Companies and the Quality, Quantity, and Cost of Physicians' Prescribing: A Systematic Review," *PLoS Medicine* 7(10): e1000352.

Steinman, M.A., Shlipak, M.G., and McPhee, S.J. (2001). "Of Principles and Pens: Attitudes and Practices of Medicine Housestaff toward Pharmaceutical Industry Promotions," *The American Journal of Medicine* 110(7): 551–557.

Sullivan, M.D., Edlund, M.J., Steffick, D., and Unützer, J. (2005). "Regular Use of Prescribed Opioids: Association with Common Psychiatric Disorders," *Pain* 119(1): 95–103.

Sunstein, C.R. (2016). "People Prefer System 2 Nudges (Kind of)," *Available at SSRN.*

Tannenbaum, D., Doctor, J.N., Persell, S.D., Friedberg, M.W., Meeker, D., Friesema, E.M., Goldstein, N.J., Linder, J.A., and Fox, C.R. (2015). "Nudging Physician Prescription Decisions by Partitioning the Order Set: Results of a Vignette-Based Study," *Journal of General Internal Medicine* 30(3): 298–304.

Thaler, R.H. and Sunstein, C.R. (2008). *Nudge: Improving Decisions About Health, Wealth, and Happiness.* London: Yale University Press.

Van Ryn, M. (2002). "Research on the Provider Contribution to Race/Ethnicity Disparities in Medical Care," *Medical Care* 40(1): I-140-I-151.

Vervloet, M., Meulepas, M.A., Cals, J.W., Eimers, M., van der Hoek, L.S., and van Dijk, L. (2016). "Reducing Antibiotic Prescriptions for Respiratory Tract Infections in Family Practice: Results of a Cluster Randomized Controlled Trial Evaluating a

Multifaceted Peer-Group-Based Intervention," *npj Primary Care Respiratory Medicine* 26: 15083.

Von Korff, M.R. (2012). "Opioids for Chronic Noncancer Pain: As the Pendulum Swings, Who Should Set Prescribing Standards for Primary Care?" *The Annals of Family Medicine* 10(4): 302–303.

Von Neumann, J. and Morgenstern, O. (2007). *Theory of Games and Economic Behavior.* Princeton, NJ: Princeton University Press.

Walsh, K.E., Adams, W.G., Bauchner, H., Vinci, R.J., Chessare, J.B., Cooper, M.R., Hebert, P.M., Schainker, E.G., and Landrigan, C.P. (2006). "Medication Errors Related to Computerized Order Entry for Children," *Pediatrics* 118(5): 1872–1879.

Wanzer, S., Federman, D.D., Adelstein, S.J., Cassel, C.K., Cassem, E.H., Cranford, R.E., Hook, E.W., Lo, B., Moertel, C.G., Safar, P., et al. (1989). "The Physician's Responsibility Towards Hopelessly Ill Patients—a Second Look," *New England Journal of Medicine* 320: 844–849.

Ward, C., Morella, L., Ashburner, J.M., and Atlas, S.J. (2014). "An Interactive, All-Payer, Multi-Domain Primary Care Performance Dashboard," *Journal of Ambulatory Care Management* 37(4): 339–348.

Wazana, A. (2000). "Physicians and the Pharmaceutical Industry: Is a Gift Ever Just a Gift?" *JAMA* 283(3): 373–380.

Weinshall-Margel, K. and Shapard, J. (2011). "Overlooked Factors in the Analysis of Parole Decisions," *Proceedings of the National Academy of Sciences* 108(42): E833–E833.

White, K.M., Hogg, M.A., and Terry, D.J. (2002). "Improving Attitude-Behavior Correspondence through Exposure to Normative Support from a Salient Ingroup," *Basic and Applied Social Psychology* 24: 91–103.

12

USING BEHAVIORAL ECONOMICS TO IMPROVE PEOPLE'S DECISIONS ABOUT PURCHASING HEALTH INSURANCE

Andrew J. Barnes, Thomas Rice, and Yaniv Hanoch

Introduction

A number of countries rely on the principles of managed competition to govern their health care systems. As devised by Alain Enthoven (1978, 1993), the key element of such a system is reliance on consumers to choose wisely among competing health insurance plans, which are subject to public regulations to help ensure that competition is based on value rather than undesirable factors such as selection of the healthiest enrollees. Ideally, individuals pick from among a number of competing plans based on such factors as premiums, perceived quality, and benefits provided. Plans compete on this basis to attract customers and therefore, to remain competitive, insurers need to provide good value. One way in which they would do so is to negotiate effectively with providers, other suppliers, and pharmaceutical companies, so that enrollees can receive these services and products most cost-efficiently.

Enthoven called his idea a "consumer choice health plan," and, as implied, it is driven by people making intelligent choices. Enthoven's ideas have had strong influence even outside of the United States, including in the United Kingdom and Sweden (Freeman and Moran, 2000). The health care system in which managed competition has been most influential is the Netherlands, which, since 2006, has adopted a system most akin to the principles espoused by Enthoven (Rosenau and Lako, 2008).

Evidence is quite mixed, however, as to whether consumers' behavior matches the theory's predictions. As discussed below, while research indicates that premiums are influential in consumers' initial choice of insurance plans, most evidence indicates a severe status quo bias: few people change plans when they are permitted to do so during annual open enrollment periods (Abaluck and Gruber, 2013; Day and Nadash, 2012; Zhou and Zhang, 2012). Moreover, consumer choice decisions on health plans often do not appear to be utility maximizing, with those of inferior value often being selected—not surprising, given further evidence of very low health insurance literacy among the general public. Moreover, many consumers do not have a good grasp of health insurance plan attributes (e.g., out-of-pocket maximums) nor how combinations of plan attributes (e.g., low premiums vs. high cost-sharing) affect the risk protection various coverage options confer, thus making health insurance purchases exceedingly complex.

Researchers have put much effort over the years into *choice architecture*, which involves tailoring information about alternatives in a way to make choices simple and comprehensible. From a behavioral economics standpoint, judicious use of choice architecture has several potential benefits:

- Reducing *information overload*, so that the choice is less complex and the decision process does not appear to be so daunting.
- Increasing *salience* by highlighting key information, and downplaying or even omitting less important information.
- Reducing *status quo bias* by making it easier to switch—that is, reducing people's *inertia*.
- Recognizing that most people are subject to *overconfidence bias*. An example from our own work is that seniors are less able to make choices that are in their best interest than are younger people—yet they are more confident in their ability to do so (Hanoch et al., 2009).

This chapter reviews some of this evidence and discusses ways in which behavioral economics can and has been used to improve health insurance choices. While we discuss various limitations and disadvantages of employing the types of choice architecture alluded to above, the chapter concludes that behavioral economics policy tools are powerful complements to standard economic policy solutions. Currently in health insurance markets, consumers often make choices that are decidedly not in their interest; choice architects have the potential to help people improve these decisions.

Background

The managed competition model is based on three key elements: (1) transparent information about benefit design and prices for a particular plan, (2) good information about all alternative choices, allowing people to compare benefits and prices between plans, and (3) rules and regulations to ensure that choices are made by consumers on these bases, rather than by insurers by designing plans to cherry-pick healthy enrollees.

The basic idea is straightforward: when both the costs (premiums, cost-sharing requirements) and benefits (covered services, provider networks) of each plan are presented to consumers, they can pick the one that provides them with the best value. Even so, it is necessary to establish certain rules that insurers must abide by—hence, why it is *managed* rather than traditional competition. Typically, insurers must offer a choice of insurance plans; if there is a subsidy provided to the consumer (or employee), it needs to be the same for all plan choices so that the person is fully responsible for the marginal expenditures of more expensive plans; everyone is charged the same amount irrespective of his or her health status and is guaranteed the right to purchase and renew coverage from any company; and objective information, consistent across all plans, must be provided. Recognizing that consumers are at a disadvantage on their own due to difficulty understanding, choosing and using health insurance coverage, a *sponsor*—typically organized by government—oversees the enrollment process to ensure that there are adequate choices and that insurers are complying with all rules. Sponsors may enact many other rules, such as whether and how plans market their products, whether plan benefits need to be standardized, and the method by which payments to health insurance plans are risk-adjusted.

While no country has applied all of the tenets of managed competition to the health insurance market, some have gone further than others. In the US, the model forms the basis of much employer-sponsored coverage, and it also underlies the Medicare Advantage (managed care) market for the publicly financed universal health insurance scheme for the elderly, blind, and disabled, as well as the Medicare prescription drug program (Part D), a standalone drug benefit that is publicly financed but privately administered. As noted, the Netherlands has probably gone further than any other country in basing its system on the tenets of managed competition. Not only does it have an active system of competing private insurers, but it has established a state-of-the-art risk adjustment system to make it difficult for insurers to compete by selecting the healthiest consumers (Van Ginneken et al., 2013).

There is little question that economic factors underlie certain choices that people make in health insurance. In particular, premiums tend to play a strong role in people's initial choice of plan, and when people use the services covered by their plan, cost-sharing requirements can have a significant impact on quelling service usage. The impact of premiums on plan choice is captured by the economic concept of the price elasticity of demand for health insurance. Researchers have conducted a number of studies on this, calculating elasticities largely in the range of -0.1 to -0.7 (Rice and Unruh, 2016), implying that as premiums rise by 10 percent, demand for an insurance plan falls by between 1 and 7 percent. This figure varies according to a number of factors, including the particular type of health insurance being considered. For example, in the US, the elasticity for Medicare beneficiaries is smaller (that is, closer to zero), in part because attachment to a particular provider can be more important than price (Buchmueller, 2006).

There are, nevertheless, two concerns about relying on consumer choice of competing insurers that emerge from the literature. The first regards the concept of *health insurance literacy*: do people understand the structure of health benefits and basic cost-sharing concepts well enough to make effective choices (Paez et al., 2014)? Second, most evidence indicates that people are very reluctant to switch to another health plan, even when it would appear to be in their economic interest to do so (Abaluck and Gruber, 2013).

More than half of the US adult population lacks numeracy skills essential to understand health insurance information (Peters et al., 2014) and the majority of consumers likely to enroll in plans sold in the *marketplaces* (also called *exchanges*) created under the Affordable Care Act (ACA) report that they are not confident that they understand basic insurance terminology (Blumberg et al., 2013). In Massachusetts, 40 percent of insurance exchange enrollees found health insurance information difficult to comprehend (Sinaiko et al., 2013). Previous studies have shown that insured people do not understand key insurance terms, risk, and the likely out-of-pocket costs when they experience an illness, nor do they understand what is and is not covered by their insurance plan (Cunningham et al., 2001; Kim et al., 2013; Paez et al., 2013). More complicated concepts, such as which providers are covered by their plans, show even lower levels of comprehension. Limited understanding of health insurance is particularly acute among low-income and otherwise disadvantaged populations (Kim et al., 2013; McCormack et al., 2009; Paez et al., 2013; Politi et al., 2013).

This lack of knowledge about benefits and coverage results in consumers frequently making poor insurance decisions (Loewenstein et al., 2013). Research comparing the quality of insurance choices of low-income, rural individuals with those of an Internet-savvy sample found that both groups had difficulty choosing insurance in a hypothetical marketplace that mimicked the federally-facilitated marketplaces, largely due to low health insurance literacy (Barnes et al., 2015).

Several studies demonstrate that poor health insurance literacy results in people making unambiguously bad choices for themselves (Bhargava et al., 2015; Sinaiko and Hirth, 2011). Specifically, researchers have examined whether employees who are offered a choice of health plans pick ones that are inferior to other choices in all important dimensions and in all possible health states—that is, they pick *dominated choices*. Evidence from the employer-sponsored market suggests that choosing dominated options, when they are available in an employee's plan choice set, is common (35% in one study and 55% in another) and leads to significant excess spending (Bhargava et al., 2015; Sinaiko and Hirth, 2011). A disproportionate number of employees making poor choices were older and had lower incomes. Results from hypothetical insurance choice experiments suggest that poor health insurance literacy was responsible for employees' "naïve considerations of health risk and price, rather than a sensible comparison of plan value" (Bhargava et al., 2015, p. 2). These findings add to a growing literature on health insurance literacy, most of which concludes that consumers do not understand key health insurance terms and have difficulty aligning what they want in an insurance plan with what they choose (Barnes et al., 2015; Loewenstein et al., 2013; McCormack et al., 2009; Paez et al., 2013).

The second concern about grounding an insurance-based system on consumer choice among insurance plans relates to status quo bias: the common observation that people are very reluctant to switch plans when they are permitted to do so during open enrollment periods. People, of course, may be happy with their plans or, even if they are not, unwilling to spend the effort to review their choices each year. Even so, switching rates are sometimes surprisingly low—a phenomenon true in European countries as well. In the United States, the area that has received the most attention is the Medicare Part D prescription drug program. Even though it is estimated that the average beneficiary could save between $300 and $400 per year by switching drug plans (Abaluck and Gruber, 2013; Zhou and Zhang, 2012), only about 10 percent of enrollees voluntarily change plans each year (Hoadley et al., 2015). Further, depending on the year, only 2–11 percent of beneficiaries are in the plan that minimizes their drug spending (Abaluck and Gruber, 2013). Part of the reason for this inertia is the sheer amount of choice: an average of almost 30 plans from which to choose in 2016 (albeit down from over 50 when the program began in 2006, after which time the federal government limited the number of different plans a particular company could sell). Similar patterns are emerging in the marketplaces arising out of the ACA. During the second annual enrollment period, about half of consumers who chose a plan in 2014 on healthcare.gov and maintained marketplace coverage in 2015 actively renewed their coverage (53%, 2.21 million; U.S. Department of Health and Human Services, 2015). The remaining 47 percent did not actively choose a plan and were automatically re-enrolled in their current plan. In regard to plan switching, in California, about 10 percent of marketplace enrollees switched plans between 2014 and 2015 (Covered California, 2015), similar to switching rates in the Part D market (Hoadley et al., 2015).

Similar patterns have been found in the Netherlands and Switzerland, two countries in which consumers are responsible for choosing among competing health care plans. Examining health insurance in Switzerland, Frank and Lamiraud (2009) have shown that while the number of health insurance choices in each Swiss canton rose from 39 in 1998 to 52 in 2003—with high price variability for similar products—the switching rate actually decreased from 4.8 percent in 1997 to 2.1 percent in 2000. A somewhat similar picture emerges from analyzing the Dutch market. The enactment of the Health Insurance Act in the Netherlands in 2006 has created a suitable environment for researchers to test consumers' sensitivity to price and quality. Analysis by Lako and Rosenau (2009) indicates several shortcomings: there has been a lack of cost control; satisfaction is low and perceived quality is down; and there is little indication that consumers are responsive to premiums. Like in Switzerland, annual switching rates are around 2 percent (Day and Nadash, 2012). While younger Dutch consumers are more sensitive to premiums, older consumers are sensitive to plan quality when deciding to switch, and searching for plan information tended to predict the probability of switching more among higher- rather than lower-educated consumers (Boonen et al., 2016).

Classic models of insurance choice are founded upon informed and rational decision-makers seeking to mitigate financial risk. Yet, consumers often employ mental shortcuts, exhibit decision biases, and lack the knowledge they are typically assumed to have on how health insurance benefits work and how to effectively use them, which results in suboptimal choices (Handel and Kolstad, 2015a). Moreover, some consumers are probably not able to make good choices, costing either themselves or the public a great deal. Zhang and colleagues (2015) found that the US Medicare program and its beneficiaries could save over $400 per capita—a total of $150 million for those with a diagnosis of schizophrenia in 2010—if it assigned such people to the Part D plan that minimized their drug spending. The following section discusses several promising approaches informed by behavioral economics to aid consumers in their understanding, evaluation, comparison, and selection of health insurance plans.

Using behaviorally informed policies to improve health insurance purchasing

The body of evidence supporting the effectiveness of behaviorally informed policies to improve health insurance purchasing is nascent but promising (Baicker et al., 2012). Much of the evidence is rooted in the Medicare Part D prescription drug coverage literature, with more recent studies examining employer-sponsored and non-group markets, including the ACA marketplaces. As discussed earlier in this chapter, consumers are afflicted by decision errors when purchasing coverage because of suboptimal search and decision strategies, limited proficiency understanding health insurance plan features and how these features affect health care utilization and spending, and poor designs in health insurance decision environments. This section reviews research on the promising approaches to applying behavioral economics to improve coverage decisions in employer-sponsored, private non-group, and public payer markets. These approaches are purposeful designs to the health plan decision environment made by choice architects who aim to improve consumer welfare by providing accurate, useful, timely information and helping consumers understand and use this information to compare, choose, and switch to better plans than they would have without these interventions (American Institutes for Research, 2015). Next, we will present common tools employed by choice architects to influence plan choice and then discuss the evidence on how the application of these tools has improved consumers' coverage decisions.

Promising approach 1: smart defaults

Evidence from the employer-sponsored market suggests that choosing dominated options, when they are available in an employee's plan choice set, is common and leads to significant excess spending (Bhargava et al., 2015; Sinaiko and Hirth, 2011). These findings are supported by experimental work that suggests that dominated choices are a result of a combination of factors including low health insurance literacy, overweighting some costs (e.g., premium, co-pay) and underweigthing others (e.g., deductible, out-of-pocket maximum) in health insurance choices, and challenges in making reasonable forecasts of expected health and health care utilization (Barnes et al., 2015; Bhargava et al., 2015; Johnson et al., 2013).

One promising approach from behavioral economics is using consumers' inertia to their benefit and defaulting them into health insurance plans that are affordable, offer adequate risk protection, and are high quality. Such *smart defaults* are increasingly pursued by choice architects to leverage behavioral insights into policies designed to mitigate pervasive decisions errors and improve consumer welfare in challenging choice contexts. Default options have been shown to substantially alter consumer choices in a number of decision environments (Thaler and Sunstein, 2008). Health insurance decision environments are ripe for the application of smart defaults, although the use of defaults in coverage choices to date is limited. Stronger smart defaults for health insurance should aim to automatically enroll consumers in the actuarially best high-quality plan given their expected spending, risk tolerance, and preferred providers and networks. In line with other such techniques, consumers could choose to opt out of plan defaults.

The Netherlands provides a possible example of applying smart defaults. Under the Dutch system, there is a mandatory annual deductible of €360 per year, but insurers also offer additional, albeit voluntary, deductibles of €100–500 in €100 increments. Van Winssen and colleagues (2015a) find that only 11 percent of the Dutch insured population chose a voluntary deductible in 2014 even though nearly half (48%) would have profited financially from doing so because they would have paid lower premiums. They attribute the low uptake to several cognitive biases: status quo bias, loss aversion, risk asymmetries between gains and losses, ambiguity aversion, and debt aversion (Van Winssen et al., 2015b). One of several policy alternatives they offer based on behavioral economics is making a voluntary deductible the default option, perhaps for groups most likely to benefit from it, such as younger people and those with a history of low utilization.

Our own work on the potential for stronger smart defaults to improve decision-making suggests that plan recommendations improve decision-making in the hypothetical experiments modeled off the ACA marketplaces (Barnes, Hanoch, and Rice, 2016). In one experiment, participants were provided with 10,000 Monopoly dollars (representing 100 actual dollars); their choice objective was to earn the most money by choosing an insurance plan that would best cover expected costs under a known probability of illness. We found that when choosing insurance without a plan recommendation, less than half of the sample chose the earnings maximizing alternative. Recommending the earnings maximizing plan, however, increased the probability of choosing the plan that minimized total expected annual spending on health by 21 percentage points after adjustment, illustrating the potential of introducing similar smart defaults into marketplace decision environments. While much of the work on defaults focuses on steering consumers toward better options, planners may wish to consider the converse, steering consumers away from worse options. For example, stronger smart defaults could require a consumer to actively choose to re-enroll in a low quality plan when less expensive, higher quality plans are available that offer drugs on a consumer's formulary and/or include providers in the consumer's current network. In fact, this is already the case for choice of private insurance plans in the US Medicare program. In the Medicare Advantage program, beneficiaries wishing to re-enroll during the annual open enrollment period must explicitly do so (that is, they are not renewed in the same plan by default) if Medicare has provided a low rating for plan quality (Rice et al., 2014).

To fully leverage the promise of smart defaults, two elements are key. First, those designing health insurance decision environments must obtain and incorporate information of sufficient quality to predict consumers' health risks and the trajectory of expected utilization and determine plan defaults (Handel and Kolstad, 2015b; Quincy, 2012). Second, plan defaults should be welfare-improving by providing more expected value than other plan options while considering consumers' risk preferences and their preference for particular providers. With the right information, smart defaults could be designed in a way to preserve freedom of choice while preventing costly decision errors (Handel and Kolstad, 2015b).

Weaker smart defaults could include designing the choice architecture to highlight plan attributes that are systematically underweighted by consumers but ultimately drive the quality of insurance choices. For example, consumers would benefit from having salient summary information on plan costs and quality including total costs based on expected utilization as well as health plan and participating provider quality measures (American Institutes for Research, 2015). Additionally, health insurance choice architects could default the sort order of plans to further emphasize the total expected costs and/or plan and participating provider quality ratings (Quincy, 2012). Such designs would highlight some of the most critical information consumers should consider when choosing a plan while allowing consumers to change the default plan filters based on other preferences and needs (Quincy, 2012).

Promising approach 2: reducing comparison frictions

Comparison frictions arise when consumers have difficulty obtaining, understanding and using information about how plan choices differ. These frictions can result in decision errors where stated preferences (what consumers say they want) and revealed preferences (what they actually choose) do not align. When comparison frictions impede consumers' ability to use information to evaluate plan options, they may lead to consumers not making any choice (i.e., status quo bias) or actively choosing a dominated option.

Promising approaches to reduce comparison frictions generally employ choice architecture to aid consumers when choosing from a menu of options. Given that many consumers have difficulty understanding health insurance terms, to reduce reliance on health insurance literacy, plain language should be used in describing plan attributes. Additionally, websites and printed insurance material should be designed to allow consumers to "scan" the information rather than requiring them to carefully read the information (American

Institutes for Research, 2015). Insurance choice architects should take advantage of inherent decision heuristics of consumers by designing shortcuts into the decision-making process including step-by-step summaries of how to choose a plan, helping consumers identify what plan attributes are most important, and grouping and summarizing information by meaningful categories (American Institutes for Research, 2015). Examples from our work and others echo these recommendations, finding that symbolic representations of price and quality aid health insurance decision-making (Barnes et al., 2012; Greene et al., 2016). In particular, we found that consumers facing symbolic rather than numeric price frames had an 8 percentage point higher probability of choosing the lowest cost prescription drug plan (Barnes et al., 2012). However, Hibbard and colleagues (2012) have found that symbolic price frames resulted in the lower likelihood of choosing high value (low cost and high access or quality) providers, suggesting perhaps that such interventions may not generalize across health insurance and health care provider shopping.

Another promising approach to reducing comparison frictions is standardization. This requires that all insurers sell policies that have exactly the same benefits, so that it is easier for consumers to make "apples to apples" comparisons between plans. Usually, there are multiple tiers of coverage (e.g., most extensive, middle, least extensive); standardization occurs within each tier. Historically, the introduction of mandatory standardization into the private market for supplemental insurance for Medicare beneficiaries ("Medigap") in the early 1990s both improved consumer understanding and reduced marketing abuses by companies and agents (Fox et al., 2003). More recently, evidence from the Massachusetts Health Connector found that standardization of plans within tiers led consumers in the state-run marketplace to select more generous health coverage and improved consumer welfare (Ericson and Starc, 2013). Standardizing plans in the ACA marketplaces also shifted the market share of plans by changing the plan attributes consumers focused on in their decisions and increasing the salience of differences in risk protection across tiers (Ericson and Starc, 2013).

Reducing comparison frictions can also result in lower status quo bias by increasing the salience of differences in total expected annual costs between insurance plans. A randomized field experiment conducted by Kling et al. (2012) found that Medicare Part D beneficiaries receiving a letter with personalized cost information about rival plans resulted in an 11 percentage point difference in the probability of switching plans and a $100 annual reduction in expected consumer costs. Experimental evidence from a national sample of nearly 7,500 non-elderly adults showed that presenting total cost information when choosing a health insurance plan increased the probability of selecting the cost minimizing plan by 3.0 to 10.6 percentage points, depending on the scenario. Importantly, younger, minority, less educated, and lower income participants and those having difficulty affording medical bills and with health insurance terminology benefited the most from these reductions in comparison frictions (Barnes et al., 2016).

Promising approach 3: fewer and better choices

There has been a growing criticism of the notion that increased choice benefits consumers. Iyengar and colleagues, for example, have shown that offering more choices—whether chocolate, jams, or job offers—can negatively affect performance and reduce satisfaction (Iyengar and Lepper, 2000; Iyengar et al., 2004; Iyengar et al., 2006). In one illustrative study, Iyengar, Huberman, and Jiang (2004) examined 80,000 individuals' willingness to join a 401(k) retirement plan in relation to the number of plans being offered. The authors found that the more options employees had, the less likely they were to join the program due to *choice overload*. Others have shown similar findings in decisions about health insurance and retirement saving plans in the US (Besedeš et al., 2012) and in other countries (Schram and Sonnemans, 2011).

Choice size has also been identified as an important factor affecting health insurance decisions. In a series of experiments, participants more easily identified the cost minimizing plans when fewer choices were available (Barnes et al., 2012, 2015; Hanoch et al., 2009; Johnson et al., 2013). In a similar line of studies, Hanoch and colleagues (2011) have shown that the number of prescription insurance plans available to consumers affected their decision strategy, or how they evaluated information and, as a consequence, their decision quality. Using Mouselab—a technique that allows researchers to follow the patterns made by movements of a computer mouse to capture the information acquisition process and gain insight on how individuals make decisions—the researchers found that an increase in plan choice size (from 3 to 9) was associated with reduction in the amount of information examined, reduction in time the information was viewed, and utilization of poorer decision strategy. Furthermore, the study also revealed that older adults were more likely to be negatively affected by increases in choice size, such that they utilized suboptimal search strategies. Interestingly, an investigation with medical trainees (Hanoch et al., 2010) reported similar trends, indicating that even highly educated and numerate health care professionals tend to employ less optimal decision strategies as the number of choices increases, and hence perform worse when faced with more Part D insurance plans. More recent experimental evidence from plan choices in hypothetical ACA marketplaces also finds that the quality of decision-making declines as the number of options increases, even when many of the options are strongly or weakly dominated (Barnes et al., 2015). These errors in judgment and decision-making are consistent with Herbert Simon's (1955) theory of bounded rationality and suggest that consumers facing information-rich environments often make suboptimal choices due to cognitive overload.

To reduce these costly decision errors, governments and employers could take a more active role in market entry and quality control of the plans insurers offer to prospective beneficiaries. First and foremost, no dominated options should be available in a consumer's choice set. Second, as has been suggested by

Abaluck and Gruber (2013), limiting choice sets to a few plans with the lowest average annual costs would likely improve consumer welfare. Another option, suggested by Rice and Cummings (2010) in the context of the Medicare prescription drug insurance market, is to have the government winnow down the number of choices to a manageable number, based on quality and price, and offer program beneficiaries this smaller list of curated options. Third, in the absence of explicitly restricting choices, pursuing other promising approaches discussed in this chapter, like reducing comparison frictions, informing consumers, and using smart defaults, may indirectly result in fewer plan offerings (Day and Nadash, 2012).

Promising approach 4: combining just-in-time education with other behaviorally informed approaches

In parallel to smart defaults, the fourth promising approach is to tailor information and educate consumers in real time. This "just-in-time" education, or information interventions timed to occur when consumers are making their purchasing decisions, aims to improve consumer health insurance literacy in the moment and in context to help them make better health insurance choices. Johnson et al. (2013) conducted a series of plan choice experiments on a sample of participants similar demographically to those choosing in the ACA marketplaces. Evidence from their hypothetical health insurance decision tasks suggests that when participants were provided with a total cost calculator to reduce comparison frictions in the first part of the experiment, they still chose the high-value plan option less than half of the time. One key explanation for this observation is that even when comparison frictions are reduced, limited understanding of insurance can impede the effectiveness of behaviorally informed policies often aimed at reducing task complexity rather than increasing knowledge. However, in the second part of the experiment, when participants were also given a just-in-time tutorial about how to calculate annual costs and answered a quiz requiring one correct choice to complete, plan choices improved noticeably. Further, when cost calculators and just-in-time education were combined with smart defaults, average participants' performance increased to be on par with financially literate MBA students.

While there is limited evidence on pairing real-time education interventions with behaviorally informed approaches, such combined interventions offer the promise of leveraging both standard economic policies to correct information gaps with behavioral economics policies that aim to overcome challenging tasks and the mental shortcuts and biases they engender in decision-makers. This implies that even though consumers may be provided with total cost estimators by their employers or the ACA marketplaces, more substantive welfare improvements arise by combining behavioral economics tools to circumvent faulty decision-making while also changing, through education, the decision-making processes consumers employ when choosing an insurance plan.

Limitations of behavioral economics in improving health insurance purchasing decisions

Providing consumers with information about health insurance plans seems to be, on the face of it, the most promising way to inform and educate consumers, and hence allow them to make the best decisions. Our review, however, challenges the idea that more information, in and of itself, is always helpful or sufficient. Rather we offer an alternative perspective and prescription, which emerge from behavioral economics theory and evidence. Nevertheless, while behaviorally informed tools are promising, they are not without limitations and criticism. For example, one limitation is that much of the research on behavioral economics and health insurance purchasing has been done using hypothetical choice experiments and as such may not fully capture the complexity of real-life decisions. Many of these hypothetical choice experiments do not include incentives, which could lead to smaller effects. However, recent experimental evidence from hypothetical insurance experiments suggests that incentives do little to improve decision-making (Johnson et al., 2013) and our own work suggests that consumers do poorly at making health insurance choices regardless of whether money is at stake (Barnes, Hanoch, and Rice, 2015, 2016). Further, actual plan choices from employer-sponsored markets suggest that as many as one-half of employees choose plans that place them at greater financial risk compared to other plans available to them (Bhargava et al., 2015; Sinaiko and Hirth, 2011).

Additionally, many consumers and policymakers will vehemently object to any attempt to curb choice. Having choice—whether in the supermarket, education or health care—is one of the fundamental pillars of both market economics and many people's views of democracy. Indeed, perhaps the most famous written work on behavioral economics, the book *Nudge* (Thaler and Sunstein, 2008), argues against reducing the quantity of choices available, but rather to steer people among the (potential) multitude of choices already in a market.

While reducing consumer choice is likely to encounter objections, policymakers and consumers alike might be more willing to accept other changes to the choice architecture. There are, nevertheless, concerns about employing choice architecture based on these behavioral economics concepts. Perhaps the main one is that by providing less detailed information, some consumers will be unable to make their best choices. There is a tradeoff between comprehensiveness and comprehensibility. Choice architects must be careful, in addition, not to go too far in "dumbing down" what is presented to consumers.

In a similar vein, if the number of choices is reduced, the right choice for one person may be unavailable. An example of this was discussed earlier: standardizing insurance benefits, a key tenet of managed competition, according to Enthoven (1993). To provide an example from the 2016 California health insurance marketplace, all Silver plans are required to have identical benefits including a $2,250 annual deductible for individuals, while Gold plans must have a $0 deductible. Deductibles are, of course, just one of many insurance benefits that are standardized in California. When considered over the large range of benefits that are offered by insurers, standardized benefits make decision-making much easier because the main attribute that consumers need to compare is the premium. Nevertheless, if one believes that a $1,000 deductible would be her best option, that choice is not available when benefits are standardized, as is the case in the California marketplace.

Further, many such changes to choice architecture—like introducing defaults or other types of nudges—have also been criticized for their paternalistic nature. They might appear, rightly or wrongly, to manipulate people into making decisions that go against their will. Finally, despite good intentions, governments and other organizations might use behavioral economic tools like defaults incorrectly, such as by assigning people to less than optimal insurance plans. In summary, despite the potential pitfalls of relying on behavioral economics to improve insurance choices, the literature reviewed above strongly suggests that the current insurance systems are far from optimal and the obstacles that consumers have encountered urgently call for improvements that take advantage of behavioral insights.

Conclusion

Health insurance choices are among the most important financial and health-related decisions that people make. Ensuring that they make the best choices is, thus, paramount to individuals and policymakers alike. While traditional economic tools and theories have made inroads toward this goal, drawing on ideas and findings from behavioral economics can result in further crucial contributions. Indeed, with many consumers choosing plans or failing to switch out of plans that offer them insufficient coverage given their expected health care needs, incorporating insights from state-of-the art research on consumer decision-making and behavioral economics has the potential to improve the quality of health insurance purchases.

Despite these well-documented, costly difficulties consumers face when purchasing health insurance, the promising approaches discussed in this chapter have yet to be widely implemented (Wong et al., 2015). Employers, government insurance program administrators, and policymakers should consider such approaches to make meaningful changes to the insurance decision environments to improve consumer welfare and better leverage plan subsidies. Given the rather new arrival of behavioral economics ideas into this market, there is a scope for many more studies utilizing different methodologies.

In some countries, governments are taking a leading role. The Behavioural Insight Team (colloquially called the "Nudge Unit") in the United Kingdom is involved in multiple policy realms to improve consumer decision-making. Following suit, the Obama Administration in the United States began its own government-sponsored research unit that relies on behavioral economics insights.

Called the Social and Behavioral Sciences Team (SBST), it released its first annual report in 2015, which focused on a number of efforts in such diverse areas as increasing employee participation in employer-sponsored retirement plans, student loan debt, and increasing the efficiency and effectiveness of government programs. One project in particular is relevant to health insurance. People who had begun, but not completed, an application for health insurance on the federal marketplace were randomly assigned to groups that received different letters to encourage completion of the application. The most effective letter resulted in a 13 percent increase in enrollment in the marketplace. That letter included both a picture of the sender and a place where the recipient was asked to write down the date by which they planned to complete the application (SBST, 2015).

There is, we believe, a wide spectrum of ideas that could be employed to help consumers make better health insurance decisions. We hope that this chapter will stimulate and inspire future research in examining different ways of presenting and simplifying choice architecture, which would help aid consumers in making these difficult, but vital, decisions.

References

Abaluck, J. and Gruber, G. (2013). "Evolving Choice Inconsistencies in Choice of Prescription Drug Insurance." National Bureau of Economic Research Working Paper 19163. Available at http://www.nber.org/papers/w19163.pdf (accessed January 3, 2017).

American Institutes for Research (2015). "Marketplace Survey Improvement Guide," Centers for Medicare and Medicaid Services, Center for Clinical Standards and Quality. Available at https://qhpcahps.cms.gov/marketplace-survey-improvement-guide (accessed January 3, 2017).

Baicker, K., Congdon, W.J., and Mullainathan, S. (2012). "Health Insurance Coverage and Take-Up: Lessons from Behavioral Economics," *Milbank Quarterly* 90(1): 107–134.

Barnes, A.J., Hanoch, Y., and Rice, T. (2015). "Determinants of Coverage Decisions in Health Insurance Marketplaces: Consumers' Decision-making Abilities and the Amount of Information in their Choice Environment," *Health Services Research* 50 (1): 58–80.

Barnes, A.J., Hanoch, Y., and Rice, T. (2016). "Can Plan Recommendations Improve the Coverage Decisions of Vulnerable Populations in Health Insurance Marketplaces?" *PLoS ONE* 11(3): doi:10.1371/journal.pone.0151095.

Barnes, A.J., Hanoch, Y., Long S.K., and Rice T. (2016). "Moving Beyond Blind Men and Elephants: Providing Total Estimated Annual Costs Improves Health Insurance Decision-making," Working Paper.

Barnes, A.J., Hanoch, Y., Wood, S., Liu, P.J., and Rice, T. (2012). "One Fish, Two Fish, Red Fish, Blue Fish: Effects of Price Frames, Brand Names, and Choice Set Size on Medicare Part D Insurance Plan Decisions," *Medical Care Research and Review* 69 (4): 460–473.

Besedeš, T., Deck, C., Sarangi, S., and Shor, M. (2012). "Age Effects and Heuristics in Decision Making," *Review of Economics and Statistics* 94(2): 580–595.

Bhargava, S., Loewenstein, G., and Sydnor, J. (2015). "Do Individuals Make Sensible Health Insurance Decisions? Evidence from a Menu with Dominated Options,"

Working Paper 21160, National Bureau of Economic Research. Available at http://www.nber.org/papers/w21160 (accessed January 3, 2017).

Blumberg, L.G., Long, S.K., Kenney, G.M., and Goin, D. (2013). "Public Understanding of Basic Health Insurance Concepts on the Eve of Health Reform." Available at http://hrms.urban.org/briefs/hrms_literacy.html (accessed January 3, 2017).

Boonen, L.H.H.M, Laske-Aldershof, T., and Schut, F.T. (2016). "Switching Health Insurers: The Role of Price, Quality, and Consumer Information Search," *European Journal of Health Economics* 17(3): 339–353.

Buchmueller, T.C. (2006). "Price and the Health Plan Choices of Retirees," *Journal of Health Economics* 25(1): 81–101.

Covered California. (2015). "Covered California Has Taken Steps to Renew 92 Percent of Enrolled Consumers." Available at http://news.coveredca.com/2015/01/covered-california-has-taken-steps-to.html (accessed April 16, 2015).

Cunningham, P.J., Denk, C., and Sinclair, M. (2001). "Do Consumers Know How Their Health Plan Works?" *Health Affairs* 20(2): 159–166.

Day, R. and Nadash, P. (2012). "New State Insurance Exchanges Should Follow the Example of Massachusetts By Simplifying Choices Among Health Plans," *Health Affairs* 31(5): 982–989.

Enthoven, A.C. (1978). "Consumer-Choice Health Plan—A National Health Insurance Proposal Based on Regulated Competition in the Private Sector," *New England Journal of Medicine* 298(13): 709–720.

Enthoven, A.C. (1993). "The History and Principles of Managed Competition," *Health Affairs* 12, Supplement: 24–48.

Ericson, K.M. and Starc, A. (2013). "How Product Standardization Affects Choice: Evidence from the Massachusetts Health Insurance Exchange," Working Paper 19527, National Bureau of Economic Research. Available at http://www.nber.org/papers/w19527 (accessed January 3, 2017).

Fox, P.D., Snyder, R.E., and Rice, T. (2003). "Medigap Reform Legislation of 1990: A 10-Year Review," *Health Care Financing Review* 24(3): 121–137.

Frank R.G. and Lamiraud, K. (2009). "Choice, Price Competition and Complexity in Markets for Health Insurance," *Journal of Economic Behavior & Organization* 71(2): 550–562.

Freeman, R. and Moran, M. (2000). "Reforming Health Care in Europe," *West European Politics* 23(2): 35–58.

Greene, J., Hibbard, J.H., and Sacks, R.M. (2016). "Summarized Costs, Placement of Quality Stars, and Other Online Displays Can Help Consumers Select High-value Plans," *Health Affairs* 35(4): 671–679.

Handel, B.R. and Kolstad, J.T. (2015a). "Health Insurance for 'Humans': Information Frictions, Plan Choice, and Consumer Welfare," *American Economic Review* 105(8): 2449–2500.

Handel, B.R. and Kolstad, J.T. (2015b). "Getting the Most from Marketplaces: Smart Policies on Health Insurance Choice," Discussion Paper 2015–08, The Hamilton Project. Available at http://www.hamiltonproject.org/assets/files/smart_policies_on_health_insurance_choice_final_proposal.pdf (accessed January 3, 2017).

Hanoch, Y., Miron-Shatz, T., Cole, H., Himmelstein, M., and Federman, A.D. (2010). "Choice, Numeracy, and Physicians-in-Training Performance: The Case of Medicare Part D." *Health Psychology* 29(4): 454–459.

Hanoch, Y., Rice, T., Cummings, J., and Wood, S. (2009). "How Much Choice Is Too Much?: The Case of the Medicare Prescription Drug Benefit," *Health Services Research* 44(4): 1157–1168.

Hanoch, Y., Wood, S., Barnes, A., Liu, P.J., and Rice, T. (2011). "Choosing the Right Medicare Prescription Drug Plan: The Effect of Age, Strategy Selection, and Choice Set Size," *Health Psychology* 30(6): 719–727.

Hibbard, J.H., Greene, J., Sofaer, S., Firminger, K., and Hirsch, J. (2012). "An Experiment Shows that a Well-designed Report on Costs and Quality Can Help Consumers Choose High-value Health Care," *Health Affairs* 31(3): 560–568.

Hoadley, J., Summer, L., Hargrave, E., Stromberg, S., Cubanski, J., and Neuman, T. (2015). "To Switch or Be Switched: Examining Changes in Drug Plan Enrollment among Medicare Part D Low-Income Subsidy Enrollees." Kaiser Family Foundation. Available at http://kff.org/medicare/report/to-switch-or-be-switched-examining-changes-in-drug-plan-enrollment-among-medicare-part-d-low-income-subsidy-enrollees/ (accessed January 3, 2017).

Iyengar, S.S. and Lepper, M.R. (2000). "When Choice is Demotivating: Can One Desire Too Much of a Good Thing?" *Journal of Personality and Social Psychology* 79(6): 995–1006.

Iyengar, S.S., Huberman, G., and Jiang, W. (2004). "How Much Choice Is Too Much: Determinants of Individual Contributions in 401K Retirement Plans," in O.S. Mitchell and S. Utkus (eds), *Pension Design and Structure: New Lessons from Behavioral Finance* (pp. 83–95). Oxford: Oxford University Press.

Iyengar, S.S., Wells, R.E., and Schwartz, B. (2006). "Doing Better but Feeling Worse. Looking for the 'Best' Job Undermines Satisfaction," *Psychological Science* 17(2): 143–150.

Johnson, E.J., Hassin, R., Baker, T., Bajger, A.T., and Treuer, G. (2013). "Can Consumers Make Affordable Care Affordable? The Value of Choice Architecture," *PLoS ONE* 8(12): e81521.

Kim, J., Braun, B., and Williams, A.D. (2013). "Understanding Health Insurance Literacy: A Literature Review," *Family and Consumer Sciences Research Journal* 42(1): 3–13.

Kling, J.R., Mullainathan, S., Shafir, E., Vermeulen, L.C., and Wrobel, M.V. (2012). "Comparison Friction: Experimental Evidence from Medicare Drug Plans," *The Quarterly Journal of Economics* 127: 199–235.

Lako, C.J. and Rosenau, P.V. (2009). "Demand-driven Care and Hospital Choice: Dutch Health Policy Toward Demand-driven Care: Results from a Survey into Hospital Choice," *Health Care Analysis* 17(1): 20–35.

Loewenstein, G., Friedman, J.Y., McGill, B., Ahmad, S., Linck, S., Sinkula, S., Beshears, J., Choi, J.J., Kolstad, J., Laibson, D., Madrian, B.C., List, J.A., and Volpp, K.G. (2013). "Consumers' Misunderstanding of Health Insurance," *Journal of Health Economics* 32: 850–862.

McCormack, L., Bann, C., Uhrig, J., Berkman, N., and Rudd, R. (2009). "Health Insurance Literacy of Older Adults," *Journal of Consumer Affairs* 43(2): 223–248.

Paez, K., Goldstein, J., and Ganachari, D. (2013). "Issue Brief: Developing a Measure of Health Insurance Literacy: Understanding Consumers' Ability to Choose and Use Insurance," American Institutes for Research. Available at http://www.air.org/sites/default/files/downloads/report/Health_Insurance_Literacy_Issues_Brief_2.25.13_0.pdf (accessed January 3, 2017).

Paez, K.A., Mallery, C.J., Noel, H., et al. (2014). "Development of the Health Insurance Literacy Measure (HILM): Conceptualizing and Measuring Consumer Ability to Choose and Use Private Health Insurance," *Journal of Health Communication* 19(2): 225–239.

Peters, E., Meilleur, L., and Tompkins, M.K. (2014). "Health Literacy and Numeracy: Workshop Summary," Roundtable on Health Literacy; Board on Population Health and Public Health Practice; Institute of Medicine. Washington (DC): National Academies Press, p. 17.

Politi, M.C., Kaphingst, K.A., Kreuter, M., Shacham, E., Lovell, M.C., and McBride, T. (2013). "Knowledge of Health Insurance Terminology and Details Among the Uninsured," *Medical Care Research and Review* 71(1): 85–98.

Quincy, L. (2012). "Choice Architecture: Design Decisions that Affect Consumers' Health Plan Choices," Kleimann Communication Group, Inc., and Consumers Union. Available at http://consumersunion.org/wp-content/uploads/2013/02/Choice_Architecture_Report.pdf (accessed January 3, 2017).

Rice, T. and Cummings, J. (2010). "Reducing the Number of Drug Plans for Seniors: A Proposal and Analysis of Three Case Studies," *Journal of Health Politics, Policy and Law* 35(6): 961–997.

Rice, T. and Unruh, L. (2016). *The Economics of Health Reconsidered.* Chicago, IL: Health Administration Press.

Rice, T., Jacobson, G., Cubanski, J., and Neuman, T. (2014). "The Private Health Insurance Choices of Medicare Beneficiaries: How Much Does Price Matter?," *Medical Care Research and Review* 71(6): 661–689.

Rosenau, P.V. and Lako, C.J. (2008). "An Experiment with Regulated Competition and Individual Mandates for Universal Health Care: The New Dutch Health Insurance System," *Journal of Health Politics, Policy, and Law* 33(6): 1031–1055.

Schram, A. and Sonnemans, J. (2011). "How Individuals Choose Health Insurance: An Experimental Analysis," *European Economic Review* 55(6): 799–819.

Simon, H.A. (1955). "A Behavioral Model of Rational Choice," *The Quarterly Journal of Economics* 69(1): 99–118.

Sinaiko, A.D. and Hirth, R.A. (2011). "Consumers, Health Insurance and Dominated Choices," *Journal of Health Economics* 30(2): 450–457.

Sinaiko, A.D., Ross-Degnan, D., Soumerai, S.B., Lieu, T., and Galbraith, A. (2013). "The Experience of Massachusetts Shows that Consumers Will Need Help in Navigating Insurance Exchanges," *Health Affairs* 32(1): 78–86.

Social and Behavioral Sciences Team (2015). *Annual Report.* Available at https://www.whitehouse.gov/sites/default/files/microsites/ostp/sbst_2015_annual_report_final_9_14_15.pdf (accessed January 3, 2017).

Thaler, R.H. and Sunstein, C.R. (2008) *Nudge: Improving Decisions about Health, Wealth, and Happiness.* New Haven, CT: Yale University Press.

U.S. Department of Health and Human Services. (2015). "Open enrollment 2015 re-enrollment snapshot." Available at http://www.hhs.gov/healthcare/facts/blog/2015/02/open-enrollment-2015-re-enrollment.html?utm_campaign=022515_hhs_healthcare_blog_reenrollment&utm_medium=email&utm_source=hhs_healthcare_blog&utm_content=022515_hhs_healthcare_blog_reenrollment_titlelink (accessed March 18, 2015).

Van Ginneken, E., Swartz, K., and Van der Wees, P. (2013). "Health Insurance Exchanges in Switzerland and the Netherlands Offer Five Key Lessons for the Operations of US Exchanges," *Health Affairs* 32(4): 744–752.

Van Winssen, K.P.M., Van Kleef, R.C., and Van de Ven, W.P.M.M. (2015a). "How Profitable Is a Voluntary Deductible in Health Insurance for the Consumer?" *Health Policy* 119: 688–695.

Van Winssen, K.P.M., Van Kleef, R.C., and Van de Ven, W.P.M.M. (2015b). "Potential Determinants of Deductible Uptake in Health Insurance: How to Increase Uptake in The Netherlands," *European Journal of Health Economics.* Published online November 27, 2015. Available at http://repub.eur.nl/pub/81529/ (accessed January 19, 2017).

Wong, C.A., Nirenburg, G., Polsky, D., Town, R., and Baker, T. (2015). "Insurance Plan Presentation and Decision Support on HealthCare.gov and State-Based Web Sites Created for the Affordable Care Act," *Annals of Internal Medicine* 163(4): 327–328.

Zhang, Y., Baik, S.H., and Newhouse, J.P. (2015). "Use of Intelligent Assignment to Medicare Part D Plans for People with Schizophrenia Could Produce Substantial Savings," *Health Affairs* 34(3): 455–460.

Zhou, C. and Zhang, Y. (2012). "The Vast Majority of Medicare Part D Beneficiaries Still Don't Choose the Cheapest Plans that Meet Their Medication Needs," *Health Affairs* 31(10): 2259–2265.

13

THE ROLE OF GOVERNMENT

How behavioral economics can inform policies to improve health behaviors

Aditi P. Sen and Richard G. Frank

Introduction

The application of government policy instruments can exert powerful influence on individuals' health-related behaviors. There are numerous circumstances where government uses laws, information, and regulation to affect individual behavior with the goal of promoting healthy behaviors. A number of such efforts have successfully changed behavior and improved health outcomes. For example, higher tobacco taxes and laws that prohibit smoking in public places have resulted in major declines in smoking rates, reduced exposure to second-hand smoke, and improved respiratory and cardiovascular health (Frazer et al., 2016; Marr and Huang, 2014). Similarly, laws requiring seatbelts, air bags, and child restraints, as well as minimum-drinking-age laws, have prevented thousands of deaths due to motor vehicle crashes (U.S. Department of Transportation, 2010).

For public policy to have the intended result—increasing healthy behaviors and reducing unhealthy ones—policies must effectively motivate individuals to change their actions. Calorie labeling cannot have the desired effect unless people actually order lower-calorie foods. Free breast cancer screening cannot be effective unless at-risk women avail themselves of the service and the list goes on. Thus, to be effective, government policy must be informed by an understanding of what drives individual behavior. Economics offers a framework for developing such understanding. The traditional economic model of individual behavior is based on the assumption of rational (expected) utility-maximizing consumers (and providers) and efficient market equilibria. Since Kenneth Arrow's (1963) seminal insights, the field of health economics has recognized unique challenges that health care markets present for developing analyses that can undergird policy.

Arrow (1963) focused largely on market failures stemming from two main classes of information problems: moral hazard and adverse selection. Moral hazard occurs due to an information imbalance where individuals have information about their actions and intentions that is private from the party that will bear the costs of those actions. In health care, individuals behave differently when they are insured (e.g., are more likely to engage in risky behavior) and use more health care than they would if they were not insured because the marginal cost of that care to the individual is lowered through insurance (Pauly, 1968). Adverse selection reflects a different type of information asymmetry—individuals have better (private) information regarding their health state than insurers and thus those who are mostly likely to benefit from insurance (e.g., who are relatively sicker) are more likely to become insured. This selection of high-risk consumers hinders the ability of insurers to spread risk across healthy and unhealthy enrollees and causes market failure. Arrow (1963) also recognized that health and health care are unique in terms of individual behavior and decision-making, with factors like trust playing important roles. Given the distinct nature of health-related behavior, government intervention may be more appropriate in this domain than in others where consumption is likely to be more standard. In contrast to other contexts, decisions regarding health and health care are often impacted by stress, misinterpretation of information, and other influences on decision-making, leading to cognitive and decision errors even when the cost of making mistakes is high. These decision-making errors sometimes result in inefficient choices, which can be referred to as behavioral market failures. Like more widely recognized market failures (e.g., incomplete markets, information asymmetry) these too provide a basis for public intervention through policy.

In order to design effective policy measures it is desirable to have a detailed understanding of how individuals actually make decisions, including cases where individual choices do not conform to notions of "rational" decision-making. In this chapter, we discuss the application of traditional economic policy to the context of healthy behaviors, its limitations in this setting, and the potential of behavioral economics to offer insights into a more nuanced model of individual behavior that can ultimately inform and improve the effectiveness of policies related to health behaviors (Frank, 2007).

Traditional economic policy and its limitations

Economic policy based on strong assumptions about rational individual behavior implies that government can influence behavior through two main channels: setting prices (à la Pigou) and providing information. Both of these approaches are widely used in the context of regulation and policy related to healthy behaviors. Prices are set through direct implementation of taxes and subsidies (e.g., taxing cigarette sales, providing free access to preventive services) and through passage of laws and regulations that have the effect of raising the price of engaging in unhealthy behaviors, such as prohibiting smoking in public places and requiring drivers and passengers in cars to wear seatbelts. Information provision occurs in a number of ways, including the publication of dietary guidelines, recommendations from the Centers for Disease Control and Prevention or the U.S. Preventive Services Task Force, and mandated calorie labeling.

The expectation that these types of traditional policies will change people's behavior is based on a model of a rational expected utility-maximizing individual. That approach presumes that consumers are able to accurately weigh the costs and benefits of the alternative choices and can make an informed consumption decision. Thus, in that context, setting "the right price" and ensuring that consumers have the necessary information to make a knowledgeable choice is enough to obtain the desired outcome. This traditional approach relies on observed consumption being a reflection of true preferences. The rationality assumption requires coherent and fixed preferences, a set preference domain, and no mistakes (Bernheim and Rangel, 2005). Further, it implicitly means that information framing, incentive design, and choice architecture (e.g., the number or ranking of choices) do not affect decisions.

There is abundant evidence showing that in many cases related to healthy behaviors, consumption behavior does not conform to the traditional conception of consumer choice and therefore demand curves do not reflect true preferences and marginal willingness-to-pay schedules. The departure from the view of the consumer as fully "rational" can be attributed to a variety of forces related to the circumstances in which health-related choices are made, including the importance of trust (in professionals and institutions), the stress and fear associated with significant illnesses, and the complexity of information and choices. Arrow (1963) suggested that inefficiencies in demand were due in part to insurance and the presence of moral hazard. There are, however, other examples of cases in which we do not observe the value-demand relationship that we would expect given a neoclassical model of behavior with moral hazard. For instance, the RAND Health Insurance Experiment showed that cost-sharing reduced the use of both high-value and low-value care (Newhouse, 1993), and more recent work on the use of deductibles has similarly shown that higher deductibles simply cause people to use less of all types of care, resulting in adverse health outcomes (Brot-Goldberg et al., 2015; Hsu et al., 2006). These examples of mis-ordering of preferences suggest that there are forces other than insurance that produce inefficient choices.

In the realm of healthy behaviors, the gap between preferences and choices may be clearest in the case of addictive goods. Bernheim and Rangel (2004) note that the evidence that addiction is unlike other consumption is manifest in a series of behavioral patterns that challenge the neoclassical economic model, including unsuccessful attempts to quit; the fact that individuals engaging in addictive behaviors admit that this behavior is a mistake; and a willingness to engage in pre-commitment to exert self-control over potential future mistakes. Similarly, there is evidence that individuals who smoke try unsuccessfully to quit and that those who want to lose weight are willing to make a pre-commitment to quit smoking or exercise or change eating habits in order to manage their tendency to make decisions that threaten their cessation or weight loss goals

(Gine et al., 2010; Royer et al., 2015). In addition to "risky behaviors," we also observe mistakes in health-related decision-making in areas such as the underuse of free preventive services, and errors in choosing health insurance that result in individuals choosing cost-ineffective or dominated plans (Abaluck and Gruber, 2011; Bhargava et al., 2015; Sinaiko and Hirth, 2011). These are described in some detail in Chapter 12 of this book.

These behavioral realities suggest that there are forces influencing behavior that are outside the standard, utility-maximizing model of individual behavior. These forces include behavioral biases, such as status quo bias, where individuals may be overly resistant to change (and/or suffer from inertia), and related endowment effects (overweighting owned goods more than other goods of similar value), as well as confirmation bias (putting more weight on evidence that is consistent with pre-existing views). Bounded rationality, the idea that individual rationality is likely to be limited by the ability to process available information, the complexity of choices, and a lack of willpower or cognitive ability, can also result in deviations from rational choices (Simon, 1955). These types of forces lead individuals to use heuristics and other methods as short cuts to cope with challenging decision circumstances rather than fully maximize utility. These behavioral market failures, like other market failures found in health care markets, can result in sub-optimal outcomes. Such failures offer a rationale for government intervention beyond those related to incomplete information, asymmetric information, and externalities that have typically been the focus of economic analysis. That means considering remedies beyond information provision and price setting that take aim at the types of mistakes attributable to behavioral biases. Regulation that is grounded in a behavioral model that recognizes behavioral market failures has the potential to increase social welfare. In the next section, we discuss how the application of insights from behavioral economics can improve our understanding of individual behavior and decision-making and, in turn, make policy more effective.

The behavioral economics approach

Accounting for behavioral market failures implies that an individual's final consumption decision may not reflect his or her true preferences, due to mistakes in the decision-making process. Therefore, the design of effective policy must take into consideration how individuals make decisions and the mistakes that may occur, and try to correct for these errors. In this context, the observed outcomes are judged in conjunction with the decision-making process. Following a tradition from public economics, we treat government as a Stackelberg leader and individuals and organizations in the health sector as followers in the policy game. This structure offers a useful frame for describing the process of considering both the decision-making process and outcomes in policymaking efforts.

The policy process as a Stackelberg game

A Stackelberg game is characterized by one dominant player that "moves" first and subordinate players that follow. Arriving at equilibrium in a Stackelberg game requires the leader to make conjectures about how the follower will respond to the leader's policy choices and then "solve" for the optimal policy (i.e., with the maximum payoff) based on the expected behavior. Therefore, having information about how individuals will respond to an attempt to correct a behavioral market failure is critical for developing effective policy on the part of the Stackelberg leader. The accurate incorporation of the follower's response into the leader's (government's) strategy is the key characteristic of the game when applying the Stackelberg model to the policy setting.

The Stackelberg set-up is a useful lens for considering the design process of policies related to healthy behaviors. As discussed earlier, these policies depend on changing individual behavior to achieve success. Thus, playing this game successfully—i.e., correctly anticipating individual responses and adjusting policy accordingly—depends on a good evaluation of individuals' responses. In turn, defining a complete and correct model of individuals' reactions necessitates a rich understanding of behavior. In particular, it calls for an understanding of how individuals deviate from the standard rational model and what implications this may have for their responses to policy. Understanding these deviations from the standard model and implications for choice behavior is a critical first step in building a more nuanced model of individual behavior that can be used as a basis for policy design.

DellaVigna (2009) outlines three ways in which individuals deviate from the standard rational model: non-standard beliefs (e.g., overconfidence, misinterpretation of information); non-standard decision-making (e.g., framing, "menu effects"); and non-standard preferences (time inconsistency and self-control problems, reference dependent-utility). Many of the resulting common decision errors can be seen as forms of "mis-weighting" information relevant to the situation (Loewenstein et al., 2013). These factors lead individuals to use "heuristics" rather than fully considering all the available options, choosing the default option or first choice in a list, and responding to the way information and incentives are framed, potentially to their detriment. Below we discuss each class of deviations from the standard model, and their implications for decisions related to healthy behaviors. In the Stackelberg game of policy design, incorporating non-standard beliefs, decision-making, and preferences would improve the definition of individual reaction functions and enable the government as first mover to maximize payoffs which, in the case of the government, are based on societal welfare.

Non-standard beliefs

The idea of non-standard beliefs is based on the fact that, in contrast to the expected utility framework, providing information may not be enough to affect behavior. The wedge between access to information and behavior arises due to information frictions and individual-level heterogeneity that mean the availability of information does not guarantee correct interpretation and application. The resulting misinterpretation of information can lead to nonstandard beliefs and systematic mistakes in decision-making. For example, even in the presence of information, individuals may underweight the potential costs of engaging in risky behaviors such as smoking or the use of other addictive substances. In the case of healthy eating, the evidence suggests that efforts to promote lower calorie orders through calorie labeling have had limited success; simply providing this information has not generally been enough to cause people to change their eating habits (Elbel et al., 2009). Beyond misinterpretation or mis-application of information, researchers have shown that individuals tend to be overconfident when predicting their own future behavior, which can result in irrational behaviors such as overpaying for annual gym memberships (DellaVigna and Malmendier, 2006).

These nonstandard beliefs can result in decisions that deviate from utility-maximizing consumption. In the Stackelberg model, it is important for policymakers to incorporate these types of behaviors—underweighting potential costs of risky behavior or overpaying for consumption of contracts, such as gym memberships, due to their value as a commitment device or overconfidence about future behavior—into individuals' reaction functions. In turn, a more precise, robust individual reaction function will feed into policymakers' objective function and improve their ability to design effective policy.

Non-standard decision-making

As discussed above, information availability does not translate directly into correct interpretation and application in consumption decisions. Further, the way that information is framed and presented, and what features are made most salient, can have major impacts on individual behavior and result in nonstandard decision-making. Studies have shown, for example, that reframing and emphasizing the salience of the consequences of not taking statins can greatly increase adherence among a population with Type II diabetes, relative to a small financial incentive or simple information provision (ideas42). Other work has shown that, in the area of health insurance choice, premium amounts are often the most salient feature on "shopping" websites such as Medicare PlanFinder and HealthCare.gov. Highlighting premiums can cause people to overlook the relative importance of other plan features (e.g., deductibles), leading to consumers choosing dominated plans and leaving "money on the table" (Friedman and Frank, 2016; Newhouse and McGuire, 2014).

"Menu effects," such as choice overload and partition-dependence, are also examples of nonstandard decision-making. In the expected utility model, more choices are always welfare-improving; in reality, however, we see evidence of decision fatigue, choice overload, and inertia when individuals face a complicated choice set which can cause them to make sub-optimal choices (Iyengar and Lepper, 2000; Loewenstein, 1999). Much of the health-related evidence comes from studies of health insurance plan choice, which have suggested that providing an increased number of plan options did not increase enrollment and, beyond a threshold, additional options actually decreased enrollment (McWilliams, 2011). Other studies have shown that people have trouble choosing the plan that would minimize their costs, and restricting their choice set would likely increase welfare (Abaluck and Gruber, 2011; Hanoch et al., 2009). On the provider side, there is indication that additional alternatives can lead to physicians being less likely to choose any treatment option (Redelmeier and Shafir, 1995) and that the overuse of medical alerts on electronic health records can cause decision fatigue among physicians (Van der Sijs et al., 2006).

Beyond the number of options presented, the way in which options are described and grouped, or "partitioned," can also impact individual choice. Individuals, for example, have a tendency to distribute scarce resources, such as money, evenly over available categories in many settings, from investment options to allocation of charitable donations to food choices (Benartzi and Thaler, 2001; Fox et al., 2005; Wansink et al., 2013; and for a review, see Johnson et al., 2012). Likewise, Wansink et al. (2013) showed that physically partitioning grocery carts into an area for fruits and vegetables versus other foods, and increasing the size of this area, can nudge people toward increased purchasing of healthy foods. Similarly, deliberate partitioning of food menus into a greater number of categories for healthy foods (e.g., separate categories for fruits and vegetables) and a smaller number for unhealthy foods (e.g., one category for all desserts) can increase purchasing of healthy foods when people attempt to spread their consumption over all categories (Fox et al., 2005).

As in the case of non-standard beliefs, incorporating non-standard decision-making into expectations regarding individual response to policy is valuable. It will be helpful to policymakers, for example, to recognize that more choices are not always better, and may in fact result in sub-optimal individual consumption. In this context, policymakers may find it important to limit the choices available to ensure that people can accurately compare options and not suffer from decision fatigue and inertia, enabling them to make the welfare-maximizing decision. Similarly, the regulators' view of individual reactions to information provision would be strengthened by incorporating the likely role of how that information is framed and what features are made particularly salient. For example, the traditional model would suggest that providing information on statin adherence is "enough" to cause the optimal behavior; however, a behavioral model would take into account framing and salience when hypothesizing about resulting behavior. As above, incorporating these behavioral features into individual reaction functions would strengthen policymakers' abilities to design policies that are likely to have the biggest payoffs.

Non-standard preferences

Finally, non-standard preferences lead to a number of behavioral deviations from the rational model. Present-biased preferences cause people to overweight immediate costs and benefits relative to those in the future (O'Donoghue and Rabin, 1999). This overweighting undermines efforts to engage in healthy behaviors (e.g., losing weight), which have short-term costs (e.g., going to the gym) and benefits only over the long term (e.g., preventing obesity and related health conditions). In these situations, informing patients of the long-run benefits of healthy behaviors is unlikely to be sufficient to outweigh the immediate benefits of many unhealthy behaviors, such as overeating and smoking. Time-inconsistent preferences can also lead to self-control and willpower problems and plans to engage in healthy behaviors in the future, which are undermined by high present discount rates (Cutler et al., 2003; Hoch and Loewenstein, 1991). There is evidence that people are aware of their time-inconsistency and willing to invest in managing it, for example, through the use of pre-commitment devices to incentivize gym-going and other healthy behaviors (DellaVigna and Malmendier, 2006).

Reference-dependent utility is another example of nonstandard preferences. Default options—which apply to individuals who do not take any action to change the setting or choice—have been shown to have a powerful effect on individual choice across a range of domains, from savings to health insurance choice to organ donation. In traditional models, defaults should have no effect on behavior if the transaction costs involved in opting out of the default are low. Substantial data, however, shows that due to a combination of inertia and interpretation of the default as the choice implicitly endorsed by experts (Choi et al., 2004; Madrian and Shea, 2001), individuals are highly likely to stick with the default option, even when superior alternatives are available. For example, research shows that people deposit a suboptimal amount into retirement savings due to a $0 default contribution (Thaler and Benartzi, 2004); fail to become organ donors due to the current "opt-in" system (Johnson and Goldstein, 2003); and eat larger portions of unhealthy foods if the large portion is the default on the menu (Giesen et al., 2013).

Relatedly, people often use related heuristics, such as choosing the first plan listed when shopping for health insurance. On many health plan "shopping" websites (e.g., Health Insurance Exchanges, Medicare PlanFinder), the first-listed plan is the plan with the lowest premium. Evidence suggests that people overweight premiums relative to other plan features (e.g., cost-sharing) when choosing a plan, potentially due at least in part to the increased salience of listing this plan first (Abaluck and Gruber, 2011; Ericson and Starc, 2012). The first plan may not be the pre-set default, but is suggestive of a related decision-making error, choosing the first option on the list, for similar reasons that lead to sticking with the default (e.g., increased salience and/or implicit recommendation). As with non-standard beliefs and decision-making, accounting for reference-dependence and present-biased preferences in individual reaction functions would improve regulators' ability to anticipate individual responses to policies and select the policy most likely to be effective.

Non-standard beliefs, decision-making, and preferences are factors that can have substantial effects on individual decision-making. Other factors, such as social influences (e.g., peer effects), can also have significant impacts on individual decision-making over healthy behaviors such as exercise, vaccination, smoking, and other risky behaviors (Carrell et al., 2011; Christakis and Fowler, 2007). Incorporating these non-standard influences into our understanding of how individuals make decisions can inform policy design by lending insight into how individuals are likely to respond to the policy. In the Stackelberg model, regulators have the opportunity to improve the design of policy, based on their expectation of individual response. Thus, policy that does not account for these types of behavioral market failures will necessarily be less effective than policy based in a model of individual behavior. In the next section, we discuss how behavioral economics can offer tools to guide policy design in the area of healthy behaviors.

Using behavioral economics to strengthen policy on healthy behaviors

Public policy based on the expected utility model of individual behavior generally affects consumption through information and prices, but policy rooted in a behavioral model recognizes that there are other channels—such as the way that information and incentives are presented and made more or less salient—through which policy can impact behavior. There are two broad approaches through which governmental policies can incorporate insights from behavioral economics to correct for behavioral market failure. In addition to these direct applications of behavioral economics to health behavior-related policy, an understanding of individual behavior can improve policy design by offering insights into how individuals will respond to these policies, as discussed above (Chetty, 2015).

Using behavioral economics to make traditional policies more effective

First, behavioral economics insights can complement traditional economic interventions (i.e., providing information or changing prices) to make these interventions more effective by countering misinformation, making information and price changes more salient, and improving the design and framing of price changes (e.g., framing as losses rather than gains) (Loewenstein and Ubel, 2010). Applications of behavioral economics across a range of behaviors, including savings, energy usage, and health plan choice, suggest ways to frame information to make it more salient and improve decision-making. For example, evidence suggests that social comparisons can be effectively used to reduce energy usage by sending households information about their electricity use relative to their neighbors' use (Allcott, 2011). Reducing the number of choices, simplifying the presentation of information, and making features such as total expected costs can counter choice overload and inertia when choosing health plans (Bhargava et al., 2015; Friedman and Frank, 2016) and thereby improve welfare (Abaluck and Gruber, 2011). Price changes can be more or less effective if framed as a subsidy versus a tax.

In the area of health behaviors, these types of information and price change presentation and framing techniques could be applied in a variety of contexts, including correcting underestimation of the costs of smoking and other unhealthy behaviors and promoting a healthy lifestyle. For example, as discussed above, researchers have used information reframing to improve adherence to statins and prevent heart disease. Improving the salience and framing of standard economic incentives by designing them with psychological factors in mind has also been shown to improve effectiveness. Even small financial incentives have been effective in promoting healthy behaviors such as smoking cessation, medication adherence, and exercise (Volpp et al., 2009; Volpp et al., 2008a, 2008b); the success of these incentives rests on incorporating behavioral factors such as probability weighting, regret aversion, mental accounting, and present-bias into their design.

New policy tools based on behavioral economics

Behavioral economics also offers a set of new policy tools (for example, changing choice architecture) that can be used to influence behavior. The behavioral economics approach to policy suggests that the very decision errors that individuals routinely make—such as relying on defaults and present-biased preferences—can even be exploited to make policy more effective (Loewenstein et al., 2007). For example, default rules or auto enrollment can be used to promote desired behaviors, including the prescription of generic versus brand drugs (Patel et al., 2014); increased participation in public health insurance programs and improved health plan choice; and increased participation in organ donor registries (Johnson and Goldstein, 2003). Decision-support tools such as cost calculators on health insurance exchanges that help consumers estimate expected spending and choose the appropriate health plan can also be informed and improved by incorporating insights from behavioral economics (Johnson et al., 2013). Behavioral economics also suggests that even simple reminders and prompts to schedule time to take specific actions can improve rates of healthy behaviors such as getting flu shots (Milkman et al., 2011) and cancer screenings (Rutter et al., 2006; Sheeran and Orbell, 2000), healthy eating (Gollwitzer and Sheeran, 2006), and wearing seatbelts (Austin et al., 2006).

Applying behavioral economics in policy

There is a growing focus on applying insights from behavioral economics to public policy. This has been manifested in many ways, including the establishment of government-associated offices in the United Kingdom (UK) and United States (US), specifically focused on using behavioral science to make policy more effective (similar offices also exist in Canada and Australia). The UK Behavioural Insights Team (UK BIT, 2013), established in 2010, was the first government institution focused on the application of behavioral sciences to public policy (BIT is now a company jointly owned by the UK government, a private charity, and its employees). Since then, the United States started the Social and Behavioral Sciences Team (SBST, 2015), based at the White House, with the similar goal of using behavioral science to make public programs and services more effective by improving participation and outcomes in these programs. Both groups have a central focus on empirical testing and evaluation of behavioral science-based program interventions and redesigns, to learn what works and what doesn't in what settings.

The UK BIT and US SBST have tested applications of behavioral sciences across a range of areas, from simplifying student financial aid to increasing on-time tax payment rates, encouraging charitable giving, driving voter registration, and improving access to credit and enrollment in retirement savings plans. In the area of health, BIT researchers found that simplifying and clarifying forms based on behavioral insights reduced prescription errors (King et al., 2014). In another study, BIT ran a large randomized controlled trial to test the effect of different messages on participation in the National Health Service Organ Donor Register. They found large effects from messages that were framed in terms of reciprocity and fairness, suggesting that if these messages were used over a full year, they would result in 96,000 additional registrations (UK BIT, 2013). The US SBST conducted a study in which they tested the effectiveness of behaviorally informed letters (e.g., including an implementation prompt and a picture of the sender) to increase health insurance enrollment and found that individuals who had started the enrollment process and then received this letter were significantly more likely to finish than a control group in which participants did not receive a letter (US SBST, 2015). Other countries have used the UK and US examples to set up similar offices; the growth in testing behavioral interventions in the field will no doubt play an important role in efforts to use behavioral science to improve policy.

The Affordable Care Act (ACA) also has many examples of how behavioral economics can be used to complement traditional economic policy. The establishment of the Health Insurance Marketplaces ("Marketplaces") is a primary example. Traditional economic policy would suggest that maximizing the number of plans available, ensuring access to information about those plans, and promoting competitive pricing would be sufficient to ensure that consumers pick the utility-maximizing plan. As has been discussed, however, there is substantial evidence that individuals make errors in health plan choice despite availability of information. The design of the Marketplaces recognizes individual limitations in interpreting and applying information by facilitating comparison shopping and making the consumer plan choice experience simpler. Insights from behavioral science are manifest in many ways on the Marketplaces, including offering standardized plans (e.g., Bronze, Silver, Gold) and, increasingly, decision support tools such as expected out-of-pocket cost calculators and easily-searchable provider directories. These features are meant to reduce the effects of choice overload and decision fatigue, and increase the salience of relevant plan features, with the goal of improving individual plan choice.

In some state-based health insurance exchanges, regulators have taken even broader actions to simplify choice, including actively limiting the number of plans (e.g., by preventing insurers from offering plans with similar features in the same area) and mandating standardized benefit designs (e.g., standard deductibles and other cost-sharing parameters) (Monahan et al., 2013). Other states are using "smart displays" to show a limited number of plan choices tailored to an individual's circumstances and, by default, rank these options based on a number of factors, including consumer preferences and total costs (Rao et al., 2015). These efforts are likely to reduce choice overload and exploit consumers' tendency to stick with recommended or default options, which has the potential to improve plan choice. Going forward, federal regulators are further improving the customer shopping experience on the Marketplaces by offering "Simple Choice Plans," with a "uniform set of features," to reduce choice overload and increase the salience of plan quality through the display of quality ratings (Counihan, 2016). Outside Marketplace design, the ACA also incorporates the power of default options by requiring that employers with over 200 employees enroll employees in health insurance automatically (allowing for opt-out).

Another example of the incorporation of behavioral economics into policy is The Behavioral Interventions to Advance Self-Sufficiency (BIAS) project, which applies behavioral insights to improve the effectiveness of large-scale social service programs supported by the Administration for Children and Families (ACF) within the U.S. Department of Health and Human Services. Through the BIAS project, ACF identifies "behavioral bottlenecks" which prevent programs from reaching their full potential, such as low rates of application and difficulty completing necessary forms or other actions, and collects qualitative and quantitative data to understand the bottlenecks and why they occur (thereby improving their understanding of individual responses to their policies). They then test targeted behavioral interventions using rigorous methods including randomized controlled trials. In this context, BIAS used reminder postcards and simplifying letters to increase applications for child support payment modifications among incarcerated parents, and increased engagement in job search activities among beneficiaries receiving Temporary Assistance for Needy Families support, using priming and other tools to overcome limited cognition (Richburg-Hayes et al., 2014). A similar approach to identifying behavioral bottlenecks and designing targeted interventions is likely to prove useful in many settings related to health behaviors, such as reducing smoking, increasing healthy eating, and improving cancer screening.

Finally, the Department of Health and Human Services is testing ways that default options in electronic medical records and e-prescribing systems can be structured so as to reduce the use of high-dose long-duration opioids for pain. A randomized experiment is underway that reduces the default drug supply from 30 days to 15 when it involves an opioid medication. This is being tested in two large urban health centers. In related work, the Food and Drug Administration is working with industry to develop user-friendly formulations of the overdose reversal drug naloxone. The formulations recognize that first responders and friends and families of addicts often face chaotic circumstances connected to an overdose that likely compromise their ability to administer a drug that was designed to be delivered by a medical professional. The new formulations are designed to be administered by a layperson without training so that the treatment can be safely and effectively administered in the settings in which addicts use drugs.

Limitations of the behavioral economics approach

Designing policy based on a model of individual behavior is not without complexity. Policy design that is intended to "correct" decision errors and redirect choices in ways that policymakers expect to align preferences and behavior does so by changing the traditional forms of policy: information provision and price-setting as well as offering new policy tools, such as the use of defaults and other forms of choice architecture. These policies may involve changing how information and choices are presented, or how subsidies and prices are designed or altered. It is important to recognize that the responses of individuals to new policies may themselves be subject to behavioral biases, which can be difficult to predict. In the context of a Stackelberg game, policymakers must anticipate individuals' response to potential policies, and, based on these expected responses, choose the optimal policy design. If, however, policymakers' expectations regarding individual behavior are incorrect or incomplete, this can result in the distortion of policy and limit its effectiveness. Forming expectations regarding individual behavior is challenging on many fronts, including due to heterogeneity of individual preferences and unanticipated forms of bias.

An incomplete model of individuals' likely response to a policy may also raise ethical concerns if the policymaker's understanding of individual preferences is misinformed. This misinformation could be due to the fact that individuals have rational reasons for engaging in the behavior targeted by the policy and the government may not in fact know what is best for them (Glaeser, 2006). Thus, policies that actively restrict choice have raised concerns, and the use of "nudges," where choice environments are designed in ways that behavioral science suggests will improve decision-making without actually restricting choice (Thaler and Sunstein, 2003), have become more common.

Nevertheless, even in cases where choice is not restricted, nudges such as the use of default options have raised ethical worries. For instance, given heterogeneity of preferences, a given default option is unlikely to be optimal for all consumers, leading to ethical concerns of choosing one default option in the presence of a known variation in likely payoffs (Smith et al., 2013). Such potential issues with default options have led to growing use of alternative approaches such as active choice, where people are not defaulted into any option but are prompted to make a choice among available options, and enhanced active choice, where the preferred alternative is highlighted. Evidence has shown that requiring an active choice can counter inertia and improve rates of healthy behaviors such as flu vaccination and medication adherence (Keller et al., 2011).

Thus, applying the behavioral model in policy settings can be limited by concerns about effectiveness and ethics. Further, it is important that behavioral economics complements traditional regulation, such as influencing relative prices through taxes, rather than substituting for it. There are likely to be cases where policy based on traditional economics may be more effective than policies based purely on behavioral economics and many situations in which a combination of traditional and behavioral interventions will be optimal (Loewenstein and Ubel, 2010). Overall, these challenges point to a need for future research to broaden and deepen our understanding of individual behavior in order to inform policy design related to healthy behaviors, including the application of behavioral insights to these policies.

Conclusion

In contexts ranging from weight loss to smoking cessation to medication adherence, observed consumption frequently does not reflect true preferences due to a variety of behavioral market failures such as nonstandard beliefs, decision-making, and preferences. These market failures present an opportunity for government intervention and a rationale for policies that attempt to help individuals by improving their choices. Further, an understanding of how people's choices systematically deviate from the rational model can improve policymakers' model of individual behavior, in turn allowing policymakers to design the most effective programs.

In this chapter, we have proposed a framework for seeing policymaking through the lens of a Stackelberg model, in which regulators take individuals' response to a policy into consideration when solving for the optimal policy. In this framework, identifying the payoff-maximizing strategy on the part of the regulator depends on an accurate understanding of individuals' reaction functions, which are incorporated into the regulator's decision. For example, to design the strongest program possible aimed at reducing smoking, policymakers must have an idea of how people will respond to related policy. Policies based on the traditional model would suggest that simply making data on the risks of smoking available would be sufficient to cause people to change behavior if they consider the risks greater than the benefits. If, however, policymakers believe that smoking is partly due to a (behavioral) failure to interpret and internalize the risks associated with tobacco, then they may require graphic warning labels even if all the data on risks is available. For this policy to be successful in getting people to refrain from or quit smoking, the policymakers' model of individual behavior (i.e., that they may not interpret and understand the data on risks fully) must be correct and complete. In this and many other settings related to healthy behaviors, the effort to describe individual reaction functions requires integrating behavioral factors into the traditional model, which assumes that people will fully understand and perfectly apply available data and act rationally, free from time inconsistent preferences, vulnerability to information framing, and reliance on heuristics and reference-dependence preferences.

There is a growing body of academic and policy work testing which types of behavioral interventions are effective in which settings; however, regulations and information campaigns would benefit from more use of small-scale experiments testing the effects of different types of information provision and likely responses to rule changes. For example, understanding how consumers and physicians alike perceive new products, like biosimilar drugs, can be informative for defining the terms of regulation. Understanding consumer perceptions of risk and the effects of information framing and highlighting the salience of various factors is likely to be important in a number of settings, from promoting medication adherence to optimal health insurance plan choice to the design of outreach efforts across health settings (e.g., contraception use to stem the health consequences of the Zika virus; reducing risky behaviors among youth). A deeper understanding of opportunities to use smart defaults, active choice, and decision support tools also has the potential to promote healthy behaviors in many contexts. Such applied behavioral economics research can have immediate impacts for the design of efficient regulations in the presence of systematic departures from fully rational decision-making.

Government regulation affects many healthy behaviors and has been critical in improving health and health care in many settings. As discussed earlier, policy in this area has been associated with reductions in car accident deaths and smoking rates, increases in vaccinations, efforts to improve the health of school lunches, and access to green spaces, recreation areas, and healthy foods. Historically, regulation has been justified by market failures in the traditional sense—facilitating the provision of public goods (e.g., parks); using mandates and taxes or subsidies to address externalities, both positive (e.g., herd immunity through vaccination) and negative (e.g., second-hand smoke); addressing lack of competition in insurance and provider markets through antitrust regulation and other efforts to define and establish the rules of the market; and investing in monitoring and appropriate incentives to prevent negative effects of information asymmetries (e.g., between doctors and patients). Further, traditional policy has, through the activities of the Food and Drug Administration and other agencies, vastly increased the amount of information available to consumers when making decisions about healthy—and unhealthy—behaviors.

In this chapter, we draw attention to behavioral market failures that arise due to factors including behavioral biases, bounded rationality, and time-inconsistent preferences. These factors lead individuals to make decision errors that result in deviations from the standard model and drive a wedge between preferences and consumption; thus, observed actions may not reflect utility-maximizing behavior on the part of the individual. These sub-optimal outcomes offer a rationale for government intervention that aims to close the gap between true preferences and consumption so that individuals are able to achieve maximum well-being.

We argue that narrowing this gap cannot occur without integrating behavioral factors into the standard model of a rational consumer, typically used in policy development. This richer understanding of individual behavior can then be used to design more effective policy with the aim of promoting healthy behaviors and overall well-being.

References

Abaluck, J. and Gruber, J. (2011). "Choice Inconsistencies among the Elderly: Evidence from Plan Choice in the Medicare Part D Program," *American Economic Review* 101(4): 1180–1210.

Allcott, H. (2011). "Social Norms and Energy Conservation," *Journal of Public Economics* 95(9–10): 1082–1095.

Arrow, K.J. (1963). "Uncertainty and the Welfare Economics of Medical Care," *American Economic Review* 53(5): 941–973.

Austin, J., Sigurdsson, S.O., and Rubin, Y.S. (2006). "An Examination of the Effects of Delayed Versus Immediate Prompts on Safety Belt Use," *Environment and Behavior* 38(1): 140–149.

Benartzi, S. and Thaler, R.H. (2001). "Naïve Diversification Strategies in Defined Contribution Savings Plans," *American Economic Review* 91(1): 79–98.

Bernheim, B.D. and Rangel, A. (2004). "Addiction and Cue-Triggered Decision Processes," *American Economic Review* 94(5): 1558–1590.

Bernheim, B.D. and Rangel, A. (2005). "Behavioral Public Economics: Welfare and Policy Analysis with Non-Standard Decision Makers," Working Paper 11518, National Bureau of Economic Research. Available at http://www.nber.org/papers/w11518.pdf (accessed January 1, 2017).

Bhargava, S., Loewenstein, G., and Sydnor, J. (2015). "Do Individuals Make Sensible Health Insurance Decisions?" Working Paper 21160, National Bureau of Economic Research. Available at http://www.nber.org/papers/w21160 (accessed January 1, 2017).

Brot-Goldberg, Z.C., Chandra, A., Handel, B.R., and Kolstad, J.T. (2015). "What Does a Deductible Do? The Impact of Cost-Sharing on Health Care Prices, Quantities, and Spending Dynamics," Working Paper 21632, National Bureau of Economic Research. Available at http://www.nber.org/papers/w21632 (accessed January 1, 2017).

Carrell, S.E., Hoekstra, M., and West, J. (2011). "Is Poor Fitness Contagious? Evidence from Randomly Assigned Friends," *Journal of Public Economics* 95(7–8): 657–663.

Chetty, R. (2015). "Behavioral Economics and Public Policy: A Pragmatic Perspective," *American Economic Review* 105(5): 1–33.

Choi, J., Laibson, D., Madrian, B., and Metrick, A. (2004) "For Better or For Worse: Default Effects and 401(k) Savings Behavior," in David A. Wise (ed.), *Perspectives in the Economics of Aging*. Chicago, IL: University of Chicago Press.

Christakis, N.A. and Fowler, J.H. (2007). "The Spread of Obesity in a Large Social Network over 32 Years," *New England Journal of Medicine* 357: 370–379.

Counihan, K. (2016). "Simplifying Choices in the Marketplace—Simple Choice Plans and Quality Star Ratings," The CMS Blog. Available at https://blog.cms.gov/2016/04/29/simplifying-choices-in-the-marketplace-simple-choice-plans-and-quality-star-ratings/ (accessed January 1, 2017).

Cutler, D.M., Glaeser, E.L., and Shapiro, J.M. (2003). "Why Have Americans Become More Obese?" *Journal of Economic Perspectives* 17(3): 93–118.

DellaVigna, S. (2009). "Psychology and Economics: Evidence from the Field," *Journal of Economic Literature* 47(2): 315–372.

DellaVigna, S. and Malmendier, U. (2006). "Paying Not to Go to the Gym," *American Economic Review* 96(3): 694–719.

Elbel, B., Kersh, R., Brescoll, V.L., and Dixon, L.B. (2009). "Calorie Labeling and Food Choices: A First Look at the Effects on Low-Income People in New York City," *Health Affairs* 28(6): 1110–1121.

Ericson, K.M. and Starc, A. (2012). "Heuristics and Heterogeneity in Health Insurance Exchanges: Evidence from the Massachusetts Connector," *American Economic Review* 102(3): 493–497.

Fox, C.R., Ratner, R.K., and Lieb, D. (2005). "How Subjective Grouping of Options Influences Choice and Allocation: Diversification Bias and the Phenomenon of Partition Dependence," *Journal of Experimental Psychology: General* 134(4): 538–551.

Frank, R.G. (2007). "Behavioral Economics and Health Economics," in P. Diamond, and H. Vartiainen (eds), *Behavioral Economics and Its Applications*. Princeton, NJ: Princeton University Press.

Frazer, K., Callinan, J.E., McHugh, J., van Baarsel, S., Clarke, A., Doherty, K., and Kelleher, C. (2016). "Legislative Smoking Bans for Reducing Harms from Second-hand Smoke Exposure, Smoking Prevalence and Tobacco Consumption," *Cochrane Database of Systematic Reviews*, Issue 2. Art. No.: CD005992.

Friedman, A. and Frank, R. (2016). "Using Behavioral Economics to Improve Plan Choice in Medicare: A Randomized Experiment," Presentation at the 6th Biennial Conference of the American Society for Health Economists: New Frontiers in Health Policy and Health Care, June 13, 2016, Philadelphia, PA. Available at https://ashecon.confex.com/ashecon/2016/webprogram/Paper4742.html (accessed January 1, 2017).

Giesen, J.C.A.H., Geyskens, K., Goukens, C., and Havermans, R.C. (2013). "Changing the Default: How to Promote Healthier Food Choices," *Appetite* 71: 475.

Gine, X., Karlan, D., and Zinman, J. (2010). "Put Your Money Where Your Butt Is: A Commitment Contract for Smoking Cessation," *American Economic Journal: Applied Economics* 2(4): 213–235.

Glaeser, E.L. (2006). "Paternalism and Psychology," *University of Chicago Law Review* 73(1): 133–156.

Gollwitzer, P.M. and Sheeran, P. (2006). "Implementation Intentions and Goal Achievements: A Meta-Analysis of Its Effects and Processes," in M.P. Zanna (ed.), *Advances in Experimental Social Psychology*. New York: Academic Press.

Hanoch, Y., Rice, T., Cummings, J., and Wood, S. (2009). "How Much Choice Is Too Much? The Case of the Medicare Prescription Drug Benefit," *Health Services Research* 44(4): 1157–1168.

Hoch, S.J. and Loewenstein, G.F. (1991). "Time-Inconsistent Preferences and Consumer Self-Control," *Journal of Consumer Research* 17(4): 492–507.

Hsu, J., Price, M., Huang, J., Brand, R., Fung, V., Hui, R., Fireman, B., Newhouse, J.P., and Selby, J.V. (2006). "Unintended Consequences of Caps on Medicare Drug Benefits," *New England Journal of Medicine* 354(22): 2349–2359.

ideas42, "Sticking to Statins," ideas42 Blog. Available at http://www.ideas42.org/blog/project/sticking-statins/ (accessed January 1, 2017).

Iyengar, S.S. and Lepper, M.R. (2000). "When Choice Is Demotivating: Can One Desire Too Much of a Good Thing?" *Journal of Personality and Social Psychology* 79(6): 995–1006.

Johnson, E.J. and Goldstein, D.G. (2003). "Do Defaults Save Lives?" *Science* 302: 1338–1339.

Johnson, E.J., Hassin, R., Baker, T., Bajger, A.T., and Treuer, G. (2013). "Can Consumers Make Affordable Care Affordable? The Value of Choice Architecture," *PLOS One* 8(12): e81521.

Johnson, E.J., Shu, S.B., Dellaert, B.G.C., Fox, C.R., Goldstein, D.G., Häubl, G., Larrick, R.P., Payne, J.W., Peters, E., Schkade, D., Wansink, B., and Weber, E.U. (2012). "Beyond Nudges: Tools of a Choice Architecture," *Marketing Letters* 23: 487–504.

Keller, P.A., Harlam, B., Loewenstein, G., and Volpp, K.G. (2011). "Enhanced Active Choice: A New Method to Motivate Behavior Change," *Journal of Consumer Psychology* 21: 376–383.

King, D., Jabbar, A., Charani, E., Bicknell, C., Wu, Z., Miller, G., Gilchrist, M., Vlaev, I., Franklin, B.D., and Darzi, A. (2014). "Redesigning the 'Choice Architecture' of Hospital Prescription Charts: A Mixed Methods Study Incorporating In Situ Simulation Testing," *BMJ Open* 4(12): e005473.

Loewenstein, G. (1999). "Is More Choice Always Better?" Social Security Brief No. 7, National Academy of Social Insurance. Available at https://www.nasi.org/research/1999/more-choice-always-better (accessed January 1, 2017).

Loewenstein, G. and Ubel, P. (2010). "Economics Behaving Badly," *New York Times*. Available at http://www.nytimes.com/2010/07/15/opinion/15loewenstein.html?_r=0 (accessed January 1, 2017).

Loewenstein, G., Brennan, T., and Volpp, K.G. (2007). "Asymmetric Paternalism to Improve Health Behaviors," *Journal of the American Medical Association* 298(20): 2415–2417.

Loewenstein, G., John, L.K., and Volpp, K.G. (2013). "Using Decision Errors to Help People Help Themselves," in E. Shafir (ed.), *The Behavioral Foundations of Public Policy*. Princeton, NJ; Oxford: Princeton University Press.

Madrian, B.C. and Shea, D.F. (2001). "The Power of Suggestion: Inertia in 401k Participation and Savings Behavior," *Quarterly Journal of Economics* 116(4): 1149–1187.

Marr, C. and Huang, C. (2014). "Higher Tobacco Taxes Can Improve Health and Raise Revenue," Center on Budget and Policy Priorities. Available at http://www.cbpp.org/sites/default/files/atoms/files/6-19-13tax.pdf (accessed January 1, 2017).

McWilliams, J.M., Afendulis, C.C., McGuire, T.M., and Landon, B.E. (2011). Complex Medicare Advantage Choices May Overwhelm Seniors—Especially Those with Impaired Decision Making. *Health Affairs*. Available at http://content.healthaffairs.org/content/early/2011/08/16/hlthaff.2011.0132.full (accessed January 1, 2017).

Milkman, K.L., Beshears, J., Choi, J.J., Laibson, D., and Madrian, B. (2011). "Using Implementation Intentions Prompts to Enhance Influenza Vaccination Rates," *Proceedings of the National Academy of Sciences* 108(26): 10415–10420.

Monahan, C.H., Dash, S.J., Lucia, K.W., and Corlette, S. (2013). "Realizing Health Reform's Potential: What States Are Doing to Simplify Health Plan Choice in the Insurance Marketplaces," Commonwealth Fund. Available at http://www.commonwealthfund.org/~/media/Files/Publications/Issue%20Brief/2013/Dec/1720_Monahan_what_states_are_doing_simplify_rb.pdf (accessed January 1, 2017).

Newhouse, J. (1993) *Free for All? Lessons from the RAND Insurance Experiment*. Cambridge, MA: Harvard University Press.

Newhouse, J.P. and McGuire, T.G. (2014). "How Successful Is Medicare Advantage?" *Milbank Quarterly* 92(2): 351–394.

O'Donoghue, T. and Rabin, M. (1999). "Doing It Now or Later," *American Economic Review*, 89(1): 103–124.

Patel, M.S., Day, S., Howell, J.T., Lautenbach, G.L., Nierman, E.H., and Volpp, K.G. (2014). "Using Default Options within the Electronic Health Record to Increase the

Prescribing of Generic Equivalent Medications: A Quasi Experimental Study," *Annals of Internal Medicine* 161: S44–62.

Pauly, M.V. (1968). "The Economics of Moral Hazard: Comment," *American Economic Review* 58(3): 531–537.

Rao, A., White, J., and Allen, K. (2015). "2016 Health Insurance Exchanges: The Good, the Bad, and the Ugly: Policy Recommendations to Improve Consumer Choices," Council for Affordable Health Coverage. Available at http://cahc.net/wp-content/uploads/2016/07/CC2016HealthInsExchangeswebv.pdf (accessed January 1, 2017).

Redelmeier, D.A. and Shafir, E. (1995). "Medical Decision Making in Situations that Offer Multiple Alternatives," *Journal of the American Medical Association* 273(4): 302–305.

Richburg-Hayes, L., Anzelone, C., Dechausay, N., Datta S., Fiorillo, A., Potok, L., Darling, M., and Balz, J. (2014). "Behavioral Economics and Social Policy: Designing Innovative Solutions for Programs Supported by the Administration for Children and Families," OPRE Report No. 2014–16a, Office of Planning, Research and Evaluation, Administration for Children and Families, U.S. Department of Health and Human Services. Available at http://www.acf.hhs.gov/sites/default/files/opre/bias_final_execsummary_opre.pdf (accessed January 1, 2017).

Royer, H., Stehr, M., and Sydnor, J. (2015). "Incentives, Commitments, and Habit Formation in Exercise: Evidence from a Field Experiment with Workers at a Fortune-500 Company," *American Economic Journal: Applied Economics* 7(3): 51–84.

Rutter, D.R., Steadman, L., and Quine, L. (2006). "An Implementation Intentions Intervention to Increase Uptake of Mammography," *Annals of Behavioral Medicine* 32(2): 127–134.

Sheeran, P. and Orbell, S. (2000). "Using Implementation Intentions to Increase Attendance for Cervical Cancer Screening," *Health Psychology* 19(3): 283–289.

Simon, H.A. (1955). "A Behavioral Model of Rational Choice," *Quarterly Journal of Economics* 69(1): 99–118.

Sinaiko, A.D. and Hirth, R.A. (2011). "Consumers, Health Insurance and Dominated Choices," *Journal of Health Economics* 30(2): 450–457.

Smith, N.C., Goldstein, D.G., and Johnson, E.J. (2013). "Choice Without Awareness: Ethical and Policy Implications of Defaults," *Journal of Public Policy & Marketing* 32(2): 159–172.

Thaler, R.H. and Benartzi, S. (2004). "Save More Tomorrow: Using Behavioral Economics to Increase Employee Saving," *Journal of Political Economy* 112(1): S164–187.

Thaler, R.H. and Sunstein, C.R. (2003). "Libertarian Paternalism," *American Economic Review* 93(2): 175–179.

U.K. Behavioural Insights Team (UK BIT) (2013). "Applying Behavioural Insights to Organ Donation." Available at http://www.behaviouralinsights.co.uk/publications/applying-behavioural-insights-to-organ-donation/ (accessed January 1, 2017).

U.S. Department of Transportation (US DOT) (2010). "Lives Saved in 2009 by Restraint Use and Minimum-Drinking-Age Laws," Traffic Safety Facts, National Highway Traffic Safety Administration. Available at https://crashstats.nhtsa.dot.gov/Api/Public/ViewPublication/811383 (accessed January 1, 2017).

U.S. Social and Behavioral Sciences Team (US SBST) (2015). "Annual Report." Available at https://sbst.gov/assets/files/2015-annual-report.pdf (accessed January 1, 2017).

Van der Sijs, H., Aarts, J., Vulto, A., and Berg, M. (2006). "Overriding of Drug Safety Alerts in Computerized Physician Order Entry," *Journal of the American Medical Informatics Association* 13(2): 138–147.

Volpp, K.G., John, L.K., Troxel, A.B., Norton, L., Fassbender, J., and Loewenstein, G. (2008b). "Financial Incentive-Based Approaches for Weight Loss," *Journal of the American Medical Association* 300(22): 2631–2637.

Volpp, K.G., Loewenstein, G., Troxel, A.B., Doshi, J., Price, M., Laskin, M., and Kimmel, S.E. (2008a). "A Test of Financial Incentives to Improve Warfarin Adherence," *BMC Health Services Research* 8: 272–277.

Volpp, K.G., Troxel, A.B., Pauly, M.V., Glick, H.A., Puig, A., Asch, D.A., Galvin, R., Zhu, J., Wan, F., DeGuzman, J., Corbett, E., Weiner, J., and Audrain-McGovern, J. (2009). "A Randomized, Controlled Trial of Financial Incentives for Smoking Cessation," *New England Journal of Medicine* 360: 699–709.

Wansink, B., Payne, C.R., Herbst, K.C., and Soman, D. (2013). "Part Carts: Assortment Allocation Cues that Increase Fruit and Vegetable Purchases," *Journal of Nutrition Education and Behavior* 45(4): S42.

INDEX